The Story of Alchemy and the Beginnings of Chemistry

M. M. Pattison Muir

Alpha Editions

This edition published in 2024

ISBN : 9789362926760

Design and Setting By
Alpha Editions
www.alphaedis.com
Email - info@alphaedis.com

As per information held with us this book is in Public Domain.
This book is a reproduction of an important historical work. Alpha Editions uses the best technology to reproduce historical work in the same manner it was first published to preserve its original nature. Any marks or number seen are left intentionally to preserve its true form.

Contents

PREFACE. .... - 1 -

CHAPTER I .... - 2 -

THE EXPLANATION OF MATERIAL
CHANGES GIVEN BY THE GREEK
THINKERS. .... - 2 -

CHAPTER II. .... - 11 -

A SKETCH OF ALCHEMICAL THEORY. .... - 11 -

CHAPTER III. .... - 18 -

THE ALCHEMICAL CONCEPTION OF
THE UNITY AND SIMPLICITY OF
NATURE. .... - 18 -

CHAPTER IV. .... - 23 -

THE ALCHEMICAL ELEMENTS AND
PRINCIPLES. .... - 23 -

CHAPTER V. .... - 31 -

THE ALCHEMICAL ESSENCE. .... - 31 -

CHAPTER VI. .... - 44 -

ALCHEMY AS AN EXPERIMENTAL
ART .... - 44 -

CHAPTER VII.	- 54 -
THE LANGUAGE OF ALCHEMY	- 54 -
CHAPTER VIII.	- 61 -
THE DEGENERACY OF ALCHEMY	- 61 -
CHAPTER IX.	- 67 -
PARACELSUS AND SOME OTHER ALCHEMISTS.	- 67 -
CHAPTER X.	- 72 -
SUMMARY OF THE ALCHEMICAL DOCTRINE.—THE REPLACEMENT OF THE THREE PRINCIPLES OF THE ALCHEMISTS BY THE SINGLE PRINCIPLE OF PHLOGISTON.	- 72 -
CHAPTER XI.	- 82 -
THE EXAMINATION OF THE PHENOMENA OF COMBUSTION.	- 82 -
CHAPTER XII.	- 91 -
THE RECOGNITION OF CHEMICAL CHANGES AS THE INTERACTIONS OF DEFINITE SUBSTANCES.	- 91 -
CHAPTER XIII.	- 95 -
THE CHEMICAL ELEMENTS CONTRASTED WITH THE ALCHEMICAL PRINCIPLES.	- 95 -

CHAPTER XIV. ....- 103 -

THE MODERN FORM OF THE
ALCHEMICAL QUEST OF THE ONE
THING. ....- 103 -

FOOTNOTES ....- 116 -

PREFACE.

The Story of Alchemy and the Beginnings of Chemistry is very interesting in itself. It is also a pregnant example of the contrast between the scientific and the emotional methods of regarding nature; and it admirably illustrates the differences between well-grounded, suggestive, hypotheses, and baseless speculations.

I have tried to tell the story so that it may be intelligible to the ordinary reader.

M.M. PATTISON MUIR.

CAMBRIDGE, November 1902.

CHAPTER I

THE EXPLANATION OF MATERIAL CHANGES GIVEN BY THE GREEK THINKERS.

For thousands of years before men had any accurate and exact knowledge of the changes of material things, they had thought about these changes, regarded them as revelations of spiritual truths, built on them theories of things in heaven and earth (and a good many things in neither), and used them in manufactures, arts, and handicrafts, especially in one very curious manufacture wherein not the thousandth fragment of a grain of the finished article was ever produced.

The accurate and systematic study of the changes which material things undergo is called chemistry; we may, perhaps, describe alchemy as the superficial, and what may be called subjective, examination of these changes, and the speculative systems, and imaginary arts and manufactures, founded on that examination.

We are assured by many old writers that Adam was the first alchemist, and we are told by one of the initiated that Adam was created on the sixth day, being the 15th of March, of the first year of the world; certainly alchemy had a long life, for chemistry did not begin until about the middle of the 18th century.

No branch of science has had so long a period of incubation as chemistry. There must be some extraordinary difficulty in the way of disentangling the steps of those changes wherein substances of one kind are produced from substances totally unlike them. To inquire how those of acute intellects and much learning regarded such occurrences in the times when man's outlook on the world was very different from what it is now, ought to be interesting, and the results of that inquiry must surely be instructive.

If the reader turns to a modern book on chemistry (for instance, The Story of the Chemical Elements, in this series), he will find, at first, superficial descriptions of special instances of those occurrences which are the subject of the chemist's study; he will learn that only certain parts of such events are dealt with in chemistry; more accurate descriptions will then be given of changes which occur in nature, or can be produced by altering the ordinary conditions, and the reader will be taught to see certain points of likeness between these changes; he will be shown how to disentangle chemical occurrences, to find their similarities and differences; and, gradually, he will feel his way to general statements, which are more or less rigorous and accurate expressions of what holds good in a large number of chemical processes; finally, he will discover that some generalisations have been made which are exact and completely accurate descriptions applicable to every case of chemical change.

But if we turn to the writings of the alchemists, we are in a different world. There is nothing even remotely resembling what one finds in a modern book on chemistry.

Here are a few quotations from alchemical writings[1]:

> "It is necessary to deprive matter of its qualities in order to draw out its soul.... Copper is like a man; it has a soul and a body ... the soul is the most subtile part ... that is to say, the tinctorial spirit. The body is the ponderable, material, terrestrial thing, endowed with a shadow.... After a series of suitable treatments copper becomes without shadow and better than gold.... The elements grow and are transmuted, because it is their qualities, not their substances which are contrary." (Stephanus of Alexandria, about 620 A.D.)

> "If we would elicit our Medecine from the precious metals, we must destroy the particular metalic form, without impairing its specific properties. The specific properties of the metal have their abode in its spiritual part, which resides in homogeneous water. Thus we must destroy the particular form of gold, and change it into its generic homogeneous water, in which the spirit of gold is preserved; this spirit afterwards restores the consistency of its water, and brings forth a new form (after the necessary putrefaction) a thousand times more perfect than the form of gold which it lost by being reincrudated." (Philalethes, 17th century.)

> "The bodily nature of things is a concealing outward vesture." (Michael Sendivogius, 17th century.)

> "Nothing of true value is located in the body of a substance, but in the virtue ... the less there is of body, the more in proportion is the virtue." (Paracelsus, 16th century.)

> "There are four elements, and each has at its centre another element which makes it what it is. These are the four pillars of the world.... It is their contrary action which

keeps up the harmony and equilibrium of the mundane machinery." (Michael Sendivogius.)

"Nature cannot work till it has been supplied with a material: the first matter is furnished by God, the second matter by the sage." (Michael Sendivogius.)

"When corruptible elements are united in a certain substance, their strife must sooner or later bring about its decomposition, which is, of course, followed by putrefaction; in putrefaction, the impure is separated from the pure; and if the pure elements are then once more joined together by the action of natural heat, a much nobler and higher form of life is produced.... If the hidden central fire, which during life was in a state of passivity, obtain the mastery, it attracts to itself all the pure elements, which are thus separated from the impure, and form the nucleus of a far purer form of life." (Michael Sendivogius.)

"Cause that which is above to be below; that which is visible to be invisible; that which is palpable to become impalpable. Again let that which is below become that which is above; let the invisible become visible, and the impalpable become palpable. Here you see the perfection of our Art, without any defect or diminution." (Basil Valentine, 15th century.)

"Think most diligently about this; often bear in mind, observe and comprehend, that all minerals and metals together, in the same time, and after the same fashion, and of one and the same principal matter, are produced and generated. That matter is no other than a mere vapour, which is extracted from the elementary earth by the superior stars, or by a sidereal distillation of the macrocosm; which sidereal hot infusion, with an airy sulphurous property, descending upon inferiors, so acts and operates as that there is implanted, spiritually and invisibly, a certain power and virtue in those metals and minerals; which fume, moreover, resolves in the earth into a certain water, wherefrom all metals are thenceforth generated and ripened to their perfection, and thence proceeds this or that metal or mineral, according as one of the three principles acquires dominion, and they have much or little of sulphur and salt, or an unequal mixture of these; whence some metals are fixed—that is, constant or stable; and some are volatile and easily changeable, as is seen in gold, silver, copper, iron, tin, and lead." (Basil Valentine.)

"To grasp the invisible elements, to attract them by their material correspondences, to control, purify, and transform them by the living power of the Spirit—this is true Alchemy." (Paracelsus.)

"Destruction perfects that which is good; for the good cannot appear on account of that which conceals it.... Each one of the visible metals is a concealment of the other six metals." (Paracelsus.)

These sayings read like sentences in a forgotten tongue.

Humboldt tells of a parrot which had lived with a tribe of American Indians, and learnt scraps of their language; the tribe totally disappeared; the parrot alone remained, and babbled words in the language which no living human being could understand.

Are the words I have quoted unintelligible, like the parrot's prating? Perhaps the language may be reconstructed; perhaps it may be found to embody something worth a hearing. Success is most likely to come by considering the growth of alchemy; by trying to find the ideas which were expressed in the strange tongue; by endeavouring to look at our surroundings as the alchemists looked at theirs.

Do what we will, we always, more or less, construct our own universe. The history of science may be described as the history of the attempts, and the failures, of men "to see things as they are." "Nothing is harder," said the Latin poet Lucretius, "than to separate manifest facts from doubtful, what straightway the mind adds on of itself."

Observations of the changes which are constantly happening in the sky, and on the earth, must have prompted men long ago to ask whether there are any limits to the changes of things around them. And this question must have become more urgent as working in metals, making colours and dyes, preparing new kinds of food and drink, producing substances with smells and tastes unlike those of familiar objects, and other pursuits like these, made men acquainted with transformations which seemed to penetrate to the very foundations of things.

Can one thing be changed into any other thing; or, are there classes of things within each of which change is possible, while the passage from one class to another is not possible? Are all the varied substances seen, tasted, handled, smelt, composed of a limited number of essentially different things; or, is each fundamentally different from every other substance? Such questions as these must have pressed for answers long ago.

Some of the Greek philosophers who lived four or five hundred years before Christ formed a theory of the transformations of matter, which is essentially the theory held by naturalists to-day.

These philosophers taught that to understand nature we must get beneath the superficial qualities of things. "According to convention," said Democritus (born 460 B.C.), "there are a sweet and a bitter, a hot and a cold, and according to convention there is colour. In truth there are atoms and a void." Those investigators attempted to connect all the differences which are observed between the qualities of things with differences of size, shape, position, and movement of atoms. They said that all things are formed by the coalescence of certain unchangeable, indestructible, and impenetrable particles which they named atoms; the total number of atoms is constant; not one of them can be destroyed, nor can one be created; when a substance ceases to exist and another is formed, the process is not a destruction of matter, it is a re-arrangement of atoms.

Only fragments of the writings of the founders of the atomic theory have come to us. The views of these philosophers are preserved, and doubtless amplified and modified, in a Latin poem, Concerning the Nature of Things, written by Lucretius, who was born a century before the beginning of our era. Let us consider the picture given in that poem of the material universe, and the method whereby the picture was produced.[2]

All knowledge, said Lucretius, is based on "the aspect and the law of nature." True knowledge can be obtained only by the use of the senses; there is no other method. "From the senses first has proceeded the knowledge of the true, and the senses cannot be refuted. Shall reason, founded on false sense, be able to contradict [the senses], wholly founded as it is on the senses? And if they are not true, then all reason as well is rendered false." The first principle in nature is asserted by Lucretius to be that "Nothing is ever gotten out of nothing." "A thing never returns to nothing, but all things after disruption go back to the first bodies of matter." If there were not imperishable seeds of things, atoms, "first-beginnings of solid singleness," then, Lucretius urges, "infinite time gone by and lapse of days must have eaten up all things that are of mortal body."

The first-beginnings, or atoms, of things were thought of by Lucretius as always moving; "there is no lowest point in the sum of the universe" where they can rest; they meet, clash, rebound, or sometimes join together into groups of atoms which move about as wholes. Change, growth, decay, formation, disruption—these are the marks of all things. "The war of first-beginnings waged from eternity is carried on with dubious issue: now here, now there, the life-bringing elements of things get the mastery, and are o'ermastered in turn; with the funeral wail blends the cry which babies raise when they enter the borders of light; and no night ever followed day, nor morning night, that heard not, mingling with the sickly infant's cries, the attendants' wailings on death and black funeral."

Lucretius pictured the atoms of things as like the things perceived by the senses; he said that atoms of different kinds have different shapes, but the number of shapes is finite, because there is a limit to the number of different things we see, smell, taste, and handle; he implies, although I do not think he definitely asserts, that all atoms of one kind are identical in every respect.

We now know that many compounds exist which are formed by the union of the same quantities by weight of the same elements, and, nevertheless, differ in properties; modern chemistry explains this fact by saying that the properties of a substance depend, not only on the kind of atoms which compose the minute particles of a compound, and the number of atoms of each kind, but also on the mode of arrangement of the atoms.[3] The same doctrine was taught by Lucretius, two thousand years ago. "It often makes a great difference," he said, "with what things, and in what positions the same first-beginnings are held in union, and what motions they mutually impart and receive." For instance, certain atoms may be so arranged at one time as to produce fire, and, at another time, the arrangement of the same atoms may be such that the result is a fir-tree. The differences between the colours of things are said by Lucretius to be due to differences in the arrangements and motions of atoms. As the colour of the sea when wind lashes it into foam is different from the colour when the waters are at rest, so do the colours of things change when the atoms whereof the things are composed change from one arrangement to another, or from sluggish movements to rapid and tumultuous motions.

Lucretius pictured a solid substance as a vast number of atoms squeezed closely together, a liquid as composed of not so many atoms less tightly packed, and a gas as a comparatively small number of atoms with considerable freedom of motion. Essentially the same picture is presented by the molecular theory of to-day.

To meet the objection that atoms are invisible, and therefore cannot exist, Lucretius enumerates many things we cannot see although we know they exist. No one doubts the existence of winds, heat, cold and smells; yet no one has seen the wind, or heat, or cold, or a smell. Clothes become moist when hung near the sea, and dry when spread in the sunshine; but no one has seen the moisture entering or leaving the clothes. A pavement trodden by many feet is worn away; but the minute particles are removed without our eyes being able to see them.

Another objector urges—"You say the atoms are always moving, yet the things we look at, which you assert to be vast numbers of moving atoms, are often motionless." Him Lucretius answers by an analogy. "And herein you need not wonder at this, that though the first-beginnings of things are all in motion, yet the sum is seen to rest in supreme repose, unless when a thing exhibits motions with its individual body. For all the nature of first things lies far away from our senses, beneath their ken; and, therefore, since they are themselves beyond what you can see, they must withdraw from sight their motion as well; and the more so, that the things which we can see do yet often conceal their motions when a great distance off. Thus, often, the woolly flocks as they crop the glad pastures on a hill, creep on whither the grass, jewelled with fresh dew, summons or invites each, and the lambs, fed to the full, gambol and playfully butt; all which objects appear to us from a distance to be blended together, and to rest like a white spot on a green hill. Again, when mighty legions fill with their movements all parts of the plains, waging the mimicry of war, the glitter lifts itself up to the sky, and the whole earth round gleams with brass, and beneath a noise is raised by the mighty tramplings of men, and the mountains, stricken by the shouting, echo the voices to the stars of heaven, and horsemen fly about, and suddenly wheeling, scour across the middle of the plains, shaking them with the vehemence of their charge. And yet there is some spot on the high hills, seen from which they appear to stand still and to rest on the plains as a bright spot."

The atomic theory of the Greek thinkers was constructed by reasoning on natural phenomena. Lucretius constantly appeals to observed facts for confirmation of his theoretical teachings, or refutation of opinions he thought erroneous. Besides giving a general mental presentation of the material universe, the theory was applied to many specific transmutations; but minute descriptions of what are now called chemical changes could not be given in terms of the theory, because no searching examination of so much as one such change had been made, nor, I think, one may say, could be made under the conditions of Greek life. More than two thousand years passed before investigators began to make accurate measurements of the quantities of the substances which take part in those changes wherein certain things seem to be destroyed and other totally different things to be produced; until accurate knowledge had been obtained of the quantities of the definite substances which interact in the transformations of matter, the atomic theory could not do more than draw the outlines of a picture of material changes.

A scientific theory has been described as "the likening of our imaginings to what we actually observe." So long as we observe only in the rough, only in a broad and general way, our imaginings must also be rough, broad, and general. It was the great glory of the Greek thinkers about natural events that their observations were accurate, on the whole, and as far as they went, and the theory they formed was based on no trivial or accidental features of the facts, but on what has proved to be the very essence of the phenomena they sought to bring into one point of view; for all the advances made in our own times in clear knowledge of the transformations of matter have been made by using, as a guide to experimental inquiries, the conception that the differences between the qualities of substances are connected with differences in the weights and movements of minute particles; and this was the central idea of the atomic theory of the Greek philosophers.

The atomic theory was used by the great physicists of the later Renaissance, by Galileo, Gassendi, Newton and others. Our own countryman, John Dalton, while trying (in the early years of the 19th century) to form a mental presentation of the atmosphere in terms of the theory of atoms, rediscovered the possibility of differences between the sizes of atoms, applied this idea to the facts concerning the quantitative compositions of compounds which had been established by others, developed a method for determining the relative weights of atoms of different kinds, and started chemistry on the course which it has followed so successfully.

Instead of blaming the Greek philosophers for lack of quantitatively accurate experimental inquiry, we should rather be full of admiring wonder at the extraordinary acuteness of their mental vision, and the soundness of their scientific spirit.

The ancient atomists distinguished the essential properties of things from their accidental features. The former cannot be removed, Lucretius said, without "utter destruction accompanying the severance"; the latter may be altered "while the nature of the thing remains unharmed." As examples of essential properties, Lucretius mentions "the weight of a stone, the heat of fire, the fluidity of water." Such things as liberty, war, slavery, riches, poverty, and the like, were accounted accidents. Time also was said to be an accident: it "exists not by itself; but simply from the things which happen, the sense apprehends what has been done in time past, as well as what is present, and what is to follow after."

As our story proceeds, we shall see that the chemists of the middle ages, the alchemists, founded their theory of material changes on the difference between a supposed essential substratum of things, and their qualities which could be taken off, they said, and put on, as clothes are removed and replaced.

How different from the clear, harmonious, orderly, Greek scheme, is any picture we can form, from such quotations as I have given from their writings, of the alchemists' conception of the world. The Greeks likened their imaginings of nature to the natural facts they observed; the alchemists created an imaginary world after their own likeness.

While Christianity was superseding the old religions, and the theological system of the Christian Church was replacing the cosmogonies of the heathen, the contrast between the power of evil and the power of good was more fully realised than in the days of the Greeks; a sharper division was drawn between this world and another world, and that other world was divided into two irreconcilable and absolutely opposite parts. Man came to be regarded as the centre of a tremendous and never-ceasing battle, urged between the powers of good and the powers of evil. The sights and sounds of nature were regarded as the vestments, or the voices, of the unseen combatants. Life was at once very real and the mere shadow of a dream. The conditions were favourable to the growth of magic; for man was regarded as the measure of the universe, the central figure in an awful tragedy.

Magic is an attempt, by thinking and speculating about what we consider must be the order of nature, to discover some means of penetrating into the secret life of natural things, of realising the hidden powers and virtues of things, grasping the concealed thread of unity which is supposed to run through all phenomena however seemingly diverse, entering into sympathy with the supposed inner oneness of life, death, the present, past, and future. Magic grows, and gathers strength, when men are sure their theory of the universe must be the one true theory, and they see only through the glasses which their theory supplies. "He who knows himself thoroughly knows God and all the mysteries of His nature," says a modern writer on magic. That saying expresses the fundamental hypothesis, and the method, of all systems of magic and mysticism. Of such systems, alchemy was one.

CHAPTER II.

A SKETCH OF ALCHEMICAL THEORY.

The system which began to be called alchemy in the 6th and 7th centuries of our era had no special name before that time, but was known as the sacred art, the divine science, the occult science, the art of Hermes.

A commentator on Aristotle, writing in the 4th century A.D., calls certain instruments used for fusion and calcination "chuika organa," that is, instruments for melting and pouring. Hence, probably, came the adjective chyic or chymic, and, at a somewhat later time, the word chemia as the name of that art which deals with calcinations, fusions, meltings, and the like. The writer of a treatise on astrology, in the 5th century, speaking of the influences of the stars on the dispositions of man, says: "If a man is born under Mercury he will give himself to astronomy; if Mars, he will follow the profession of arms; if Saturn, he will devote himself to the science of alchemy (Scientia alchemiae)." The word alchemia which appears in this treatise, was formed by prefixing the Arabic al (meaning the) to chemia, a word, as we have seen, of Greek origin.

It is the growth, development, and transformation into chemistry, of this alchemia which we have to consider.

Alchemy, that is, the art of melting, pouring, and transforming, must necessarily pay much attention to working with crucibles, furnaces, alembics, and other vessels wherein things are fused, distilled, calcined, and dissolved. The old drawings of alchemical operations show us men busy calcining, cohobating, distilling, dissolving, digesting, and performing other processes of like character to these.

The alchemists could not be accused of laziness or aversion to work in their laboratories. Paracelsus (16th century) says of them: "They are not given to idleness, nor go in a proud habit, or plush and velvet garments, often showing their rings on their fingers, or wearing swords with silver hilts by their sides, or fine and gay gloves on their hands; but diligently follow their labours, sweating whole days and nights by their furnaces. They do not spend their time abroad for recreation, but take delight in their laboratories. They put their fingers among coals, into clay and filth, not into gold rings. They are sooty and black, like smiths and miners, and do not pride themselves upon clean and beautiful faces."

In these respects the chemist of to-day faithfully follows the practice of the alchemists who were his predecessors. You can nose a chemist in a crowd by the smell of the laboratory which hangs about him; you can pick him out by the stains on his hands and clothes. He also "takes delight in his laboratory"; he does not always "pride himself on a clean and beautiful face"; he "sweats whole days and nights by his furnace."

Why does the chemist toil so eagerly? Why did the alchemists so untiringly pursue their quest? I think it is not unfair to say: the chemist experiments in order that he "may liken his imaginings to the facts which he observes"; the alchemist toiled that he might liken the facts which he observed to his imaginings. The difference may be put in another way by saying: the chemist's object is to discover "how changes happen in combinations of the unchanging"; the alchemist's endeavour was to prove the truth of his fundamental assertion, "that every substance contains undeveloped resources and potentialities, and can be brought outward and forward into perfection."

Looking around him, and observing the changes of things, the alchemist was deeply impressed by the growth and modification of plants and animals; he argued that minerals and metals also grow, change, develop. He said in effect: "Nature is one, there must be unity in all the diversity I see. When a grain of corn falls into the earth it dies, but this dying is the first step towards a new life; the dead seed is changed into the living plant. So it must be with all other things in nature: the mineral, or the metal, seems dead when it is buried in the earth, but, in reality, it is growing, changing, and becoming more perfect." The perfection of the seed is the plant. What is the perfection of the common metals? "Evidently," the alchemist replied, "the perfect metal is gold; the common metals are trying to become gold." "Gold is the intention of Nature in regard to all metals," said an alchemical writer. Plants are preserved by the preservation of their seed. "In like manner," the alchemist's argument proceeded, "there must be a seed in metals which is their essence; if I can separate the seed and bring it under the proper conditions, I can cause it to grow into the perfect metal." "Animal life, and human life also," we may suppose the alchemist saying, "are continued by the same method as that whereby the life of plants is continued; all life springs from seed; the seed is fructified by the union of the male and the female; in metals also there must be the two characters; the union of these is needed for the production of new metals; the conjoining of metals must go before the birth of the perfect metal."

"Now," we may suppose the argument to proceed, "now, the passage from the imperfect to the more perfect is not easy. It is harder to practise virtue than to acquiesce in vice; virtue comes not naturally to man; that he may gain the higher life, he must be helped by grace. Therefore, the task of exalting the purer metals into the perfect gold, of developing the lower order into the higher, is not easy. If Nature does this, she does it slowly and painfully; if the exaltation of the common metals to a higher plane is to be effected rapidly, it can be done only by the help of man."

So far as I can judge from their writings, the argument of the alchemists may be rendered by some such form as the foregoing. A careful examination of the alchemical argument shows that it rests on a (supposed) intimate knowledge of nature's plan of working, and the certainty that simplicity is the essential mark of that plan.

That the alchemists were satisfied of the great simplicity of nature, and their own knowledge of the ways of nature's work, is apparent from their writings.

The author of The New Chemical Light (17th century) says: "Simplicity is the seal of truth.... Nature is wonderfully simple, and the characteristic mark of a childlike simplicity is stamped upon all that is true and noble in Nature." In another place the same author says: "Nature is one, true, simple, self-contained, created of God, and informed with a certain universal spirit." The same author, Michael Sendivogius, remarks: "It may be asked how I come to have this knowledge about heavenly things which are far removed beyond human ken. My answer is that the sages have been taught by God that this natural world is only an image and material copy of a heavenly and spiritual pattern; that the very existence of this world is based upon the reality of its heavenly archetype.... Thus the sage sees heaven reflected in Nature as in a mirror, and he pursues this Art, not for the sake of gold or silver, but for the love of the knowledge which it reveals."

The Only True Way advises all who wish to become true alchemists to leave the circuitous paths of pretended philosophers, and to follow nature, which is simple; the complicated processes described in books are said to be the traps laid by the "cunning sophists" to catch the unwary.

In A Catechism of Alchemy, Paracelsus asks: "What road should the philosopher follow?" He answers, "That exactly which was followed by the Great Architect of the Universe in the creation of the world."

One might suppose it would be easier, and perhaps more profitable, to examine, observe, and experiment, than to turn one's eyes inwards with the hope of discovering exactly "the road followed by the Great Architect of the Universe in the creation of the world." But the alchemical method found it easier to begin by introspection. The alchemist spun his universe from his own ideas of order, symmetry, and simplicity, as the spider spins her web from her own substance.

A favourite saying of the alchemists was, "What is above is as what is below." In one of its aspects this saying meant, "processes happen within the earth like those which occur on the earth; minerals and metals live, as animals and plants live; all pass through corruption towards perfection." In another aspect the saying meant "the human being is the world in miniature; as is the microcosm, so is the macrocosm; to know oneself is to know all the world."

Every man knows he ought to try to rise to better things, and many men endeavour to do what they know they ought to do; therefore, he who feels sure that all nature is fashioned after the image of man, projects his own ideas of progress, development, virtue, matter and spirit, on to nature outside himself; and, as a matter of course, this kind of naturalist uses the same language when he is speaking of the changes of material things as he employs to express the changes of his mental states, his hopes, fears, aspirations, and struggles.

The language of the alchemists was, therefore, rich in such expressions as these; "the elements are to be so conjoined that the nobler and fuller life may be produced"; "our arcanum is gold exalted to the highest degree of perfection to which the combined action of nature and art can develop it."

Such commingling of ethical and physical ideas, such application of moral conceptions to material phenomena, was characteristic of the alchemical method of regarding nature. The necessary results were; great confusion of thought, much mystification of ideas, and a superabundance of views about natural events.

When the author of The Metamorphosis of Metals was seeking for an argument in favour of his view, that water is the source and primal element of all things, he found what he sought in the Biblical text: "In the beginning the spirit of God moved upon the face of the waters." Similarly, the author of The Sodic Hydrolith clenches his argument in favour of the existence of the Philosopher's Stone, by the quotation: "Therefore, thus saith the Lord; behold I lay in Zion for a foundation a Stone, a tried Stone, a precious corner Stone, a sure foundation. He that has it shall not be confounded." This author works out in detail an analogy between the functions and virtues of the Stone, and the story of man's fall and redemption, as set forth in the Old and New Testaments. The same author speaks of "Satan, that grim pseudo-alchemist."

That the attribution, by the alchemists, of moral virtues and vices to natural things was in keeping with some deep-seated tendency of human nature, is shown by the persistence of some of their methods of stating the properties of substances: we still speak of "perfect and imperfect gases," "noble and base metals," "good and bad conductors of electricity," and "laws governing natural phenomena."

Convinced of the simplicity of nature, certain that all natural events follow one course, sure that this course was known to them and was represented by the growth of plants and animals, the alchemists set themselves the task, firstly, of proving by observations and experiments that their view of natural occurrences was correct; and, secondly, of discovering and gaining possession of the instrument whereby nature effects her transmutations and perfects her operations. The mastery of this instrument would give them power to change any metal into gold, the cure of all diseases, and the happiness which must come from the practical knowledge of the supreme secret of nature.

The central quest of alchemy was the quest of an undefined and undefinable something wherein was supposed to be contained all the powers and potencies of life, and whatever makes life worth living.

The names given to this mystical something were as many as the properties which were assigned to it. It was called the one thing, the essence, the philosopher's stone, the stone of wisdom, the heavenly balm, the divine water, the virgin water, the carbuncle of the sun, the old dragon, the lion, the basilisk, the phœnix; and many other names were given to it.

We may come near to expressing the alchemist's view of the essential character of the object of their search by naming it the soul of all things. "Alchemy," a modern writer says, "is the science of the soul of all things."

The essence was supposed to have a material form, an ethereal or middle nature, and an immaterial or spiritual life.

No one might hope to make this essence from any one substance, because, as one of the alchemists says, "It is the attribute of God alone to make one out of one; you must produce one thing out of two by natural generation." The alchemists did not pretend to create gold, but only to produce it from other things.

The author of A Brief Guide to the Celestial Ruby says: "We do not, as is sometimes said, profess to create gold and silver, but only to find an agent which ... is capable of entering into an intimate and maturing union with the Mercury of the base metals." And again: "Our Art ... only arrogates to itself the power of developing, through the removal of all defects and superfluities, the golden nature which the baser metals possess." Bonus, in his tract on The New Pearl of Great Price (16th century), says: "The Art of Alchemy ... does not create metals, or even develop them out of the metallic first-substance; it only takes up the unfinished handicraft of Nature and completes it.... Nature has only left a comparatively small thing for the artist to do—the completion of that which she has already begun."

If the essence were ever attained, it would be by following the course which nature follows in producing the perfect plant from the imperfect seed, by discovering and separating the seed of metals, and bringing that seed under the conditions which alone are suitable for its growth. Metals must have seed, the alchemists said, for it would be absurd to suppose they have none. "What prerogative have vegetables above metals," exclaims one of them, "that God should give seed to the one and withhold it from the other? Are not metals as much in His sight as trees?"

As metals, then, possess seed, it is evident how this seed is to be made active; the seed of a plant is quickened by descending into the earth, therefore the seed of metals must be destroyed before it becomes life-producing. "The processes of our art must begin with dissolution of gold; they must terminate in a restoration of the essential quality of gold." "Gold does not easily give up its nature, and will fight for its life; but our agent is strong enough to overcome and kill it, and then it also has power to restore it to life, and to change the lifeless remains into a new and pure body."

The application of the doctrine of the existence of seed in metals led to the performance of many experiments, and, hence, to the accumulation of a considerable body of facts established by experimental inquiries. The belief of the alchemists that all natural events are connected by a hidden thread, that everything has an influence on other things, that "what is above is as what is below," constrained them to place stress on the supposed connexion between the planets and the metals, and to further their metallic transformations by performing them at times when certain planets were in conjunction. The seven principal planets and the seven principal metals were called by the same names: Sol (gold), Luna (silver), Saturn (lead), Jupiter (tin), Mars (iron), Venus (copper), and Mercury (mercury). The author of The New Chemical Light taught that one metal could be propagated from another only in the order of superiority of the planets. He placed the seven planets in the following descending order: Saturn, Jupiter, Mars, Sol, Venus, Mercury, Luna. "The virtues of the planets descend," he said, "but do not ascend"; it is easy to change Mars (iron) into Venus (copper), for instance, but Venus cannot be transformed into Mars.

Although the alchemists regarded everything as influencing, and influenced by, other things, they were persuaded that the greatest effects are produced on a substance by substances of like nature with itself. Hence, most of them taught that the seed of metals will be obtained by operations with metals, not by the action on metals of things of animal or vegetable origin. Each class of substances, they said, has a life, or spirit (an essential character, we might say) of its own. "The life of sulphur," Paracelsus said, "is a combustible, ill-smelling, fatness.... The life of gems and corals is mere colour.... The life of water is its flowing.... The life of fire is air." Grant an attraction of like to like, and the reason becomes apparent for such directions as these: "Nothing heterogeneous must be introduced into our magistery"; "Everything should be made to act on that which is like it, and then Nature will perform her duty."

Although each class of substances was said by the alchemists to have its own particular character, or life, nevertheless they taught that there is a deep-seated likeness between all things, inasmuch as the power of the essence, or the one thing, is so great that under its influence different things are produced from the same origin, and different things are caused to pass into and become the same thing. In The New Chemical Light it is said: "While the seed of all things is one, it is made to generate a great variety of things."

It is not easy now—it could not have been easy at any time—to give clear and exact meanings to the doctrines of the alchemists, or the directions they gave for performing the operations necessary for the production of the object of their search. And the difficulty is much increased when we are told that "The Sage jealously conceals [his knowledge] from the sinner and the scornful, lest the mysteries of heaven should be laid bare to the vulgar gaze." We almost despair when an alchemical writer assures us that the Sages "Set pen to paper for the express purpose of concealing their meaning. The sense of a whole passage is often hopelessly obscured by the addition or omission of one little word, for instance the addition of the word not in the wrong place." Another writer says: "The Sages are in the habit of using words which may convey either a true or a false impression; the former to their own disciples and children, the latter to the ignorant, the foolish, and the unworthy." Sometimes, after descriptions of processes couched in strange and mystical language, the writer will add, "If you cannot perceive what you ought to understand herein, you should not devote yourself to the study of philosophy." Philalethes, in his Brief Guide to the Celestial Ruby, seems to feel some pity for his readers; after describing what he calls "the generic homogeneous water of gold," he says: "If you wish for a more particular description of our water, I am impelled by motives of charity to tell you that it is living, flexible, clear, nitid, white as snow, hot, humid, airy, vaporous, and digestive."

Alchemy began by asserting that nature must be simple; it assumed that a knowledge of the plan and method of natural occurrences is to be obtained by thinking; and it used analogy as the guide in applying this knowledge of nature's design to particular events, especially the analogy, assumed by alchemy to exist, between material phenomena and human emotions.

CHAPTER III.

THE ALCHEMICAL CONCEPTION OF THE UNITY AND SIMPLICITY OF NATURE.

In the preceding chapter I have referred to the frequent use made by the alchemists of their supposition that nature follows the same plan, or at any rate a very similar plan, in all her processes. If this supposition is accepted, the primary business of an investigator of nature is to trace likenesses and analogies between what seem on the surface to be dissimilar and unconnected events. As this idea, and this practice, were the foundations whereon the superstructure of alchemy was raised, I think it is important to amplify them more fully than I have done already.

Mention is made in many alchemical writings of a mythical personage named Hermes Trismegistus, who is said to have lived a little later than the time of Moses. Representations of Hermes Trismegistus are found on ancient Egyptian monuments. We are told that Alexander the Great found his tomb near Hebron; and that the tomb contained a slab of emerald whereon thirteen sentences were written. The eighth sentence is rendered in many alchemical books as follows:

"Ascend with the greatest sagacity from the earth to heaven, and then again descend to the earth, and unite together the powers of things superior and things inferior. Thus you will obtain the glory of the whole world, and obscurity will fly away from you."

This sentence evidently teaches the unity of things in heaven and things on earth, and asserts the possibility of gaining, not merely a theoretical, but also a practical, knowledge of the essential characters of all things. Moreover, the sentence implies that this fruitful knowledge is to be obtained by examining nature, using as guide the fundamental similarity supposed to exist between things above and things beneath.

The alchemical writers constantly harp on this theme: follow nature; provided you never lose the clue, which is simplicity and similarity.

The author of The Only Way (1677) beseeches his readers "to enlist under the standard of that method which proceeds in strict obedience to the teaching of nature ... in short, the method which nature herself pursues in the bowels of the earth."

The alchemists tell us not to expect much help from books and written directions. When one of them has said all he can say, he adds—"The question is whether even this book will convey any information to one before whom the writings of the Sages and the open book of Nature are exhibited in vain." Another tells his readers the only thing for them is "to beseech God to give you the real philosophical temper, and to open your eyes to the facts of nature; thus alone will you reach the coveted goal."

"Follow nature" is sound advice. But, nature was to be followed with eyes closed save to one vision, and the vision was to be seen before the following began.

The alchemists' general conception of nature led them to assign to every substance a condition or state natural to it, and wherein alone it could be said to be as it was designed to be. Each substance, they taught, could be caused to leave its natural state only by violent, or non-natural, means, and any substance which had been driven from its natural condition by violence was ready, and even eager, to return to the condition consonant with its nature.

Thus Norton, in his Ordinal of Alchemy, says: "Metals are generated in the earth, for above ground they are subject to rust; hence above ground is the place of corruption of metals, and of their gradual destruction. The cause which we assign to this fact is that above ground they are not in their proper element, and an unnatural position is destructive to natural objects, as we see, for instance, that fishes die when they are taken out of the water; and as it is natural for men, beasts, and birds to live in the air, so stones and metals are naturally generated under the earth."

In his New Pearl of Great Price (16th century), Bonus says:—"The object of Nature in all things is to introduce into each substance the form which properly belongs to it; and this is also the design of our Art."

This view assumed the knowledge of the natural conditions of the substances wherewith experiments were performed. It supposed that man could act as a guide, to bring back to its natural condition a substance which had been removed from that condition, either by violent processes of nature, or by man's device. The alchemist regarded himself as an arbiter in questions concerning the natural condition of each substance he dealt with. He thought he could say, "this substance ought to be thus, or thus," "that substance is constrained, thwarted, hindered from becoming what nature meant it to be."

In Ben Jonson's play called The Alchemist, Subtle (who is the alchemist of the play) says, " ... metals would be gold if they had time."

The alchemist not only attributed ethical qualities to material things, he also became the guardian and guide of the moral practices of these things. He thought himself able to recall the erring metal to the path of metalline virtue, to lead the extravagant mineral back to the moral home-life from which it had been seduced, to show the doubting and vacillating salt what it was ignorantly seeking, and to help it to find the unrealised object of its search. The alchemist acted as a sort of conscience to the metals, minerals, salts, and other substances he submitted to the processes of his laboratory. He treated them as a wise physician might treat an ignorant and somewhat refractory patient. "I know what you want better than you do," he seems often to be saying to the metals he is calcining, separating, joining and subliming.

But the ignorant alchemist was not always thanked for his treatment. Sometimes the patient rebelled. For instance, Michael Sendivogius, in his tract, The New Chemical Light drawn from the Fountain of Nature and of Manual Experience (17th century), recounts a dialogue between Mercury, the Alchemist, and Nature.

"On a certain bright morning a number of Alchemists met together in a meadow, and consulted as to the best way of preparing the Philosopher's Stone.... Most of them agreed that Mercury was the first substance. Others said, no, it was sulphur, or something else.... Just as the dispute began to run high, there arose a violent wind, which dispersed the Alchemists into all the different countries of the world; and as they had arrived at no conclusion, each one went on seeking the Philosopher's Stone in his own old way, this one expecting to find it in one substance, and that in another, so that the search has continued without intermission even unto this day. One of them, however, had at least got the idea into his head that Mercury was the substance of the Stone, and determined to concentrate all his efforts on the chemical preparation of Mercury.... He took common Mercury and began to work with it. He placed it in a glass vessel over the fire, when it, of course, evaporated. So in his ignorance he struck his wife, and said: 'No one but you has entered my laboratory; you must have taken my Mercury out of the vessel.' The woman, with tears, protested her innocence. The Alchemist put some more Mercury into the vessel.... The Mercury rose to the top of the vessel in vaporous steam. Then the Alchemist was full of joy, because he remembered that the first substance of the Stone is described by the Sages as volatile; and he thought that now at last he must be on the right track. He now began to subject the Mercury to all sorts of chemical processes, to sublime it, and to calcine it with all manner of things, with salts, sulphur, metals, minerals, blood, hair, aqua fortis, herbs, urine, and vinegar.... Everything he could think of was tried; but without producing the desired effect." The Alchemist then despaired; after a dream, wherein an old man came and talked with him about the "Mercury of the Sages," the Alchemist thought he would charm the Mercury, and so he used a form of incantation. The Mercury suddenly began to speak, and asked the Alchemist why he had troubled him so much, and so on. The Alchemist replied, and questioned the Mercury. The Mercury makes fun of the philosopher. Then the Alchemist again torments the Mercury by heating him with all manner of horrible things. At last Mercury calls in the aid of Nature, who soundly rates the philosopher, tells him he is grossly ignorant, and ends by saying: "The best thing you can do is to give yourself up to the king's officers, who will quickly put an end to you and your philosophy."

As long as men were fully persuaded that they knew the plan whereon the world was framed, that it was possible for them to follow exactly "the road which was followed by the Great Architect of the Universe in the creation of the world," a real knowledge of natural events was impossible; for every attempt to penetrate nature's secrets presupposed a knowledge of the essential characteristics of that which was to be investigated. But genuine knowledge begins when the investigator admits that he must learn of nature, not nature of him. It might be truly said of one who held the alchemical conception of nature that "his foible was omniscience"; and omniscience negatives the attainment of knowledge.

The alchemical notion of a natural state as proper to each substance was vigorously combated by the Honourable Robert Boyle (born 1626, died 1691), a man of singularly clear and penetrative intellect. In A Paradox of the Natural and Supernatural States of Bodies, Especially of the Air, Boyle says:—"I know that not only in living, but even in inanimate, bodies, of which alone I here discourse, men have universally admitted the famous distinction between the natural and preternatural, or violent state of bodies, and do daily, without the least scruple, found upon it hypotheses and ratiocinations, as if it were most certain that what they call nature had purposely formed bodies in such a determinate state, and were always watchful that they should not by any external violence be put out of it. But notwithstanding so general a consent of men in this point, I confess, I cannot yet be satisfied about it in the sense wherein it is wont to be taken. It is not, that I believe, that there is no sense in which, or in the account upon which, a body may he said to be in its natural state; but that I think the common distinction of a natural and violent state of bodies has not been clearly explained and considerately settled, and both is not well grounded, and is oftentimes ill applied. For when I consider that whatever state a body be put into, or kept in, it obtains or retains that state, assenting to the catholic laws of nature, I cannot think it fit to deny that in this sense the body proposed is in a natural state; but then, upon the same ground, it will he hard to deny but that those bodies which are said to be in a violent state may also be in a natural one, since the violence they are presumed to suffer from outward agents is likewise exercised no otherwise than according to the established laws of universal nature."

There must be something very fascinating and comforting in the alchemical view of nature, as a harmony constructed on one simple plan, which can be grasped as a whole, and also in its details, by the introspective processes of the human intellect; for that conception prevails to-day among those who have not investigated natural occurrences for themselves. The alchemical view of nature still forms the foundation of systems of ethics, of philosophy, of art. It appeals to the innate desire of man to make himself the measure of all things. It is so easy, so authoritative, apparently so satisfactory. No amount of thinking and reasoning will ever demonstrate its falsity. It can be conquered only by a patient, unbiassed, searching examination of some limited portion of natural events.

CHAPTER IV.

THE ALCHEMICAL ELEMENTS AND PRINCIPLES.

The alchemists were sure that the intention of nature regarding metals was that they should become gold, for gold was considered to be the most perfect metal, and nature, they said, evidently strains after perfection. The alchemist found that metals were worn away, eaten through, broken, and finally caused to disappear, by many acid and acrid liquids which he prepared from mineral substances. But gold resisted the attacks of these liquids; it was not changed by heat, nor was it affected by sulphur, a substance which changed limpid, running mercury into an inert, black solid. Hence, gold was more perfect in the alchemical scale than any other metal.

Since gold was considered to be the most perfect metal, it was self-evident to the alchemical mind that nature must form gold slowly in the earth, must transmute gradually the inferior metals into gold.

"The only thing that distinguishes one metal from another," writes an alchemist who went under the name of Philalethes, "is its degree of maturity, which is, of course, greatest in the most precious metals; the difference between gold and lead is not one of substance, but of digestion; in the baser metal the coction has not been such as to purge out its metallic impurities. If by any means this superfluous impure matter could be organically removed from the baser metals, they would become gold and silver. So miners tell us that lead has in many cases developed into silver in the bowels of the earth, and we contend that the same effect is produced in a much shorter time by means of our Art."

Stories were told about the finding of gold in deserted mines which had been worked out long before; these stories were supposed to prove that gold was bred in the earth. The facts that pieces of silver were found in tin and lead mines, and gold was found in silver mines, were adduced as proofs that, as the author of The New Pearl of Great Price says, "Nature is continually at work changing other metals into gold, because, though in a certain sense they are complete in themselves, they have not yet reached the highest perfection of which they are capable, and to which nature has destined them." What nature did in the earth man could accomplish in the workshop. For is not man the crown of the world, the masterpiece of nature, the flower of the universe; was he not given dominion over all things when the world was created?

In asserting that the baser metals could be transmuted into gold, and in attempting to effect this transmutation, the alchemist was not acting on a vague; haphazard surmise; he was pursuing a policy dictated by his conception of the order of nature; he was following the method which he conceived to be that used by nature herself. The transmutation of metals was part and parcel of a system of natural philosophy. If this transmutation were impossible, the alchemical scheme of things would be destroyed, the believer in the transmutation would be left without a sense of order in the material universe. And, moreover, the alchemist's conception of an orderly material universe was so intimately connected with his ideas of morality and religion, that to disprove the possibility of the great transmutation would be to remove not only the basis of his system of material things, but the foundations of his system of ethics also. To take away his belief in the possibility of changing other metals into gold would be to convert the alchemist into an atheist.

How, then, was the transmutation to be accomplished? Evidently by the method whereby nature brings to perfection other living things; for the alchemist's belief in the simplicity and unity of nature compelled him to regard metals as living things.

Plants are improved by appropriate culture, by digging and enriching the soil, by judicious selection of seed; animals are improved by careful breeding. By similar processes metals will be encouraged and helped towards perfection. The perfect state of gold will not be reached at a bound; it will be gained gradually. Many partial purifications will be needed. As Subtle says in The Alchemist—

'twere absurd To think that nature in the earth bred gold Perfect in the instant; something went before, There must be remote matter.... Nature doth first beget the imperfect, then Proceeds she to the perfect.

At this stage the alchemical argument becomes very ultra-physical. It may, perhaps, be rendered somewhat as follows:—

Man is the most perfect of animals; in man there is a union of three parts, these are body, soul, and spirit. Metals also may be said to have a body, a soul, and a spirit; there is a specific bodily, or material, form belonging to each metal; there is a metalline soul characteristic of this or that class of metals; there is a spirit, or inner immaterial potency, which is the very essence of all metals.

The soul and spirit of man are clogged by his body. If the spiritual nature is to become the dominating partner, the body must be mortified: the alchemists, of course, used this kind of imagery, and it was very real to them. In like manner the spirit of metals will be laid bare and enabled to exercise its transforming influences, only when the material form of the individual metal has been destroyed. The first thing to do, then, is to strip off and cast aside those properties of metals which appeal to the senses.

"It is necessary to deprive matter of its qualities in order to draw out its soul," said Stephanus of Alexandria in the 7th century; and in the 17th century Paracelsus said, "Nothing of true value is located in the body of a substance, but in the virtue ... the less there is of body the more in proportion is the virtue."

But the possession of the soul of metals is not the final stage: mastery of the soul may mean the power of transmuting a metal into another like itself; it will not suffice for the great transmutation, for in that process a metal becomes gold, the one and only perfect metal. Hence the soul also must be removed, in order that the spirit, the essence, the kernel, may be obtained.

And as it is with metals, so, the alchemists argued, it is with all things. There are a few Principles which may be thought of as conditioning the specific bodily and material forms of things; beneath these, there are certain Elements which are common to many things whose principles are not the same; and, hidden by the wrappings of elements and principles, there is the one Essence, the spirit, the mystic uniting bond, the final goal of the philosopher.

I propose in this chapter to try to analyse the alchemical conceptions of Elements and Principles, and in the next chapter to attempt some kind of description of the Essence.

In his Tract Concerning the Great Stone of the Ancient Sages, Basil Valentine speaks of the "three Principles," salt, sulphur, and mercury, the source of which is the Elements.

"There are four Elements, and each has at its centre another element which makes it what it is. These are the four pillars of the earth."

Of the element Earth, he says:—"In this element the other three, especially fire, are latent.... It is gross and porous, specifically heavy, but naturally light.... It receives all that the other three project into it, conscientiously conceals what it should hide, and brings to light that which it should manifest.... Outwardly it is visible and fixed, inwardly it is invisible and volatile."

Of the element Water, Basil Valentine says:—"Outwardly it is volatile, inwardly it is fixed, cold, and humid.... It is the solvent of the world, and exists in three degrees of excellence: the pure, the purer, and the purest. Of its purest substance the heavens were created; of that which is less pure the atmospheric air was formed; that which is simply pure remains in its proper sphere where ... it is guardian of all subtle substances here below."

Concerning the element Air, he writes:—"The most noble Element of Air ... is volatile, but may be fixed, and when fixed renders all bodies penetrable.... It is nobler than Earth or Water.... It nourishes, impregnates, conserves the other elements."

Finally, of the element Fire:—"Fire is the purest and noblest of all Elements, full of adhesive unctuous corrosiveness, penetrant, digestive, inwardly fixed, hot and dry, outwardly visible, and tempered by the earth.... This Element is the most passive of all, and resembles a chariot; when it is drawn, it moves; when it is not drawn, it stands still."

Basil Valentine then tells his readers that Adam was compounded of the four pure Elements, but after his expulsion from Paradise he became subject to the various impurities of the animal creation. "The pure Elements of his creation were gradually mingled and infected with the corruptible elements of the outer world, and thus his body became more and more gross, and liable, through its grossness, to natural decay and death." The process of degeneration was slow at first, but "as time went on, the seed out of which men were generated became more and more infected with perishable elements. The continued use of corruptible food rendered their bodies more and more gross; and human life was soon reduced to a very brief span."

Basil Valentine then deals with the formation of the three Principles of things, by the mutual action of the four Elements. Fire acting on Air produced Sulphur; Air acting on Water produced Mercury; Water acting on Earth produced Salt. Earth having nothing to act on produced nothing, but became the nurse of the three Principles. "The three Principles," he says, "are necessary because they are the immediate substance of metals. The remoter substance of metals is the four elements, but no one can produce anything out of them but God; and even God makes nothing of them but these three Principles."

To endeavour to obtain the four pure Elements is a hopeless task. But the Sage has the three Principles at hand. "The artist should determine which of the three Principles he is seeking, and should assist it so that it may overcome its contrary." "The art consists in an even mingling of the virtues of the Elements; in the natural equilibrium of the hot, the dry, the cold, and the moist."

The account of the Elements given by Philalethes differs from that of Basil Valentine.

Philalethes enumerates three Elements only: Air, Water, and Earth. Things are not formed by the mixture of these Elements, for "dissimilar things can never really unite." By analysing the properties of the three Elements, Philalethes reduced them finally to one, namely, Water. "Water," he says, "is the first principle of all things." "Earth is the fundamental Element in which all bodies grow and are preserved. Air is the medium into which they grow, and by means of which the celestial virtues are communicated to them."

According to Philalethes, Mercury is the most important of the three Principles. Although gold is formed by the aid of Mercury, it is only when Mercury has been matured, developed, and perfected, that it is able to transmute inferior metals into gold. The essential thing to do is, therefore, to find an agent which will bring about the maturing and perfecting of Mercury. This agent, Philalethes calls "Our divine Arcanum."

Although it appears to me impossible to translate the sayings of the alchemists concerning Elements and Principles into expressions which shall have definite and exact meanings for us to-day, still we may, perhaps, get an inkling of the meaning of such sentences as those I have quoted from Basil Valentine and Philalethes.

Take the terms Fire and Water. In former times all liquid substances were supposed to be liquid because they possessed something in common; this hypothetical something was called the Element, Water. Similarly, the view prevailed until comparatively recent times, that burning substances burn because of the presence in them of a hypothetical imponderable fluid, called "Caloric"; the alchemists preferred to call this indefinable something an Element, and to name it Fire.

We are accustomed to-day to use the words fire and water with different meanings, according to the ideas we wish to express. When we say "do not touch the fire," or "put your hand into the water," we are regarding fire and water as material things; when we say "the house is on fire," or speak of "a diamond of the first water," we are thinking of the condition or state of a burning body, or of a substance as transparent as water. When we say "put out the fire," or "his heart became as water," we are referring to the act of burning, or are using an image which likens the thing spoken of to a substance in the act of liquefying.

As we do to-day, so the alchemists did before us; they used the words fire and water to express different ideas.

Such terms as hardness, softness, coldness, toughness, and the like, are employed for the purpose of bringing together into one point of view different things which are alike in, at least, one respect. Hard things may differ in size, weight, shape, colour, texture, &c. A soft thing may weigh the same as a hard thing; both may have the same colour or the same size, or be at the same temperature, and so on. By classing together various things as hard or soft, or smooth or rough, we eliminate (for the time) all the properties wherein the things differ, and regard them only as having one property in common. The words hardness, softness, &c., are useful class-marks.

Similarly the alchemical Elements and Principles were useful class-marks.

We must not suppose that when the alchemists spoke of certain things as formed from, or by the union of, the same Elements or the same Principles, they meant that these things contained a common substance. Their Elements and Principles were not thought of as substances, at least not in the modern meaning of the expression, a substance; they were qualities only.

If we think of the alchemical elements earth, air, fire, and water, as general expressions of what seemed to the alchemists the most important properties of all substances, we may be able to attach some kind of meaning to the sayings of Basil Valentine, which I have quoted. For instance, when that alchemist tells us, "Fire is the most passive of all elements, and resembles a chariot; when it is drawn, it moves; when it is not drawn, it stands still"—we may suppose he meant to express the fact that a vast number of substances can be burnt, and that combustion does not begin of itself, but requires an external agency to start it.

Unfortunately, most of the terms which the alchemists used to designate their Elements and Principles are terms which are now employed to designate specific substances. The word fire is still employed rather as a quality of many things under special conditions, than as a specific substance; but earth, water, air, salt, sulphur, and mercury, are to-day the names applied to certain groups of properties, each of which is different from all other groups of properties, and is, therefore, called, in ordinary speech, a definite kind of matter.

As knowledge became more accurate and more concentrated, the words sulphur, salt, mercury, &c., began to be applied to distinct substances, and as these terms were still employed in their alchemical sense as compendious expressions for certain qualities common to great classes of substances, much confusion arose. Kunckel, the discoverer of phosphorus, who lived between 1630 and 1702, complained of the alchemists' habit of giving different names to the same substance, and the same name to different substances. "The sulphur of one," he says, "is not the sulphur of another, to the great injury of science. To that one replies that everyone is perfectly free to baptise his infant as he pleases. Granted. You may if you like call an ass an ox, but you will never make anyone believe that your ox is an ass." Boyle is very severe on the vague and loose use of words practised by so many writers of his time. In The Sceptical Chymist (published 1678-9) he says: "If judicious men, skilled in chymical affairs, shall once agree to write clearly and plainly of them, and thereby keep men from being stunned, as it were, or imposed upon by dark and empty words; it is to be hoped that these [other] men finding, that they can no longer write impertinently and absurdly, without being laughed at for doing so, will be reduced either to write nothing, or books that may teach us something, and not rob men, as formerly, of invaluable time; and so ceasing to trouble the world with riddles or impertinences, we shall either by their books receive an advantage, or by their silence escape an inconvenience."

Most of the alchemists taught that the elements produced what they called seed, by their mutual reactions, and the principles matured this seed and brought it to perfection. They supposed that each class, or kind, of things had its own seed, and that to obtain the seed was to have the power of producing the things which sprung from that seed.

Some of them, however, asserted that all things come from a common seed, and that the nature of the products of this seed is conditioned by the circumstances under which it is caused to develop.

Thus Michael Sendivogius writes as follows in The New Chemical Light, drawn from the fountain of Nature and of Manual Experience (17th century):—

> "Wherever there is seed, Nature will work through it, whether it be good or bad." "The four Elements, by their continued action, project a constant supply of seed to the centre of the earth, where it is digested, and whence it proceeds again in generative motions. Now the centre of the earth is a certain void place where nothing is at rest, and upon the margin or circumference of this centre the four Elements project their qualities.... The magnetic force of our earth-centre attracts to itself as much as is needed of the cognate seminal substance, while that which cannot be used for vital generation is thrust forth in the shape of stones and other rubbish. This is the fountain-head of all things terrestrial. Let us illustrate the matter by supposing a glass of water to be set in the middle of a table, round the margin of which are placed little heaps of salt, and of powders of different colours. If the water be poured out, it will run all over the table in divergent rivulets, and will become salt where it touches the salt, red where it touches

the red powder, and so on. The water does not change the 'places,' but the several 'places' differentiate the water.[4] In the same way, the seed which is the product of the four Elements is projected in all directions from the earth-centre, and produces different things, according to the quality of the different places. Thus, while the seed of all things is one, it is made to generate a great variety of things.... So long as Nature's seed remains in the centre it can indifferently produce a tree or a metal, a herb or a stone, and in like manner, according to the purity of the place, it will produce what is less or more pure."

CHAPTER V.

THE ALCHEMICAL ESSENCE.

In the last chapter I tried to describe the alchemical view of the interdependence of different substances. Taking for granted the tripartite nature of man, the co-existence in him of body, soul, and spirit (no one of which was defined), the alchemists concluded that all things are formed as man is formed; that in everything there is a specific bodily form, some portion of soul, and a dash of spirit. I considered the term soul to be the alchemical name for the properties common to a class of substances, and the term spirit to mean the property which was thought by the alchemists to be common to all things.

The alchemists considered it possible to arrange all substances in four general classes, the marks whereof were expressed by the terms hot, cold, moist, and dry; they thought of these properties as typified by what they called the four Elements—fire, air, water, and earth. Everything, they taught, was produced from the four Elements, not immediately, but through the mediation of the three Principles—mercury, sulphur, and salt. These Principles were regarded as the tools put into the hands of him who desired to effect the transmutation of one substance into another. The Principles were not thought of as definite substances, nor as properties of this or that specified substance; they were considered to be the characteristic properties of large classes of substances.

The chemist of to-day places many compounds in the same class because all are acids, because all react similarly under similar conditions. It used to be said that every acid possesses more or less of the principle of acidity. Lavoisier changed the language whereby certain facts concerning acids were expressed. He thought that experiments proved all acids to be compounds of the element oxygen; and for many years after Lavoisier, the alchemical expression the principle of acidity was superseded by the word oxygen. Although Lavoisier recognised that not every compound of oxygen is an acid, he taught that every acid is a compound of oxygen. We know now that many acids are not compounds of oxygen, but we have not yet sufficient knowledge to frame a complete definition of the term acid. Nevertheless it is convenient, indeed it is necessary, to place together many compounds which react similarly under certain defined conditions, and to give a common name to them all. The alchemists also classified substances, but their classification was necessarily more vague than ours; and they necessarily expressed their reasons for putting different substances in the same class in a language which arose out of the general conceptions of natural phenomena which prevailed in their time.

The primary classification of substances made by the alchemists was expressed by saying; these substances are rich in the principle sulphur, those contain much of the principle mercury, and this class is marked by the preponderance of the principle salt. The secondary classification of the alchemists was expressed by saying; this class is characterised by dryness, that by moisture, another by coldness, and a fourth by hotness; the dry substances contain much of the element Earth, the moist substances are rich in the element Water, in the cold substances the element Air preponderates, and the hot substances contain more of the element Fire than of the other elements.

The alchemists went a step further in their classification of things. They asserted that there is One Thing present in all things; that everything is a vehicle for the more or less perfect exhibition of the properties of the One Thing; that there is a Primal Element common to all substances. The final aim of alchemy was to obtain the One Thing, the Primal Element, the Soul of all Things, so purified, not only from all specific substances, but also from all admixture of the four Elements and the three Principles, as to make possible the accomplishment of any transmutation by the use of it.

If a person ignorant of its powers were to obtain the Essence, he might work vast havoc and cause enormous confusion; it was necessary, therefore, to know the conditions under which the potencies of the Essence became active. Hence there was need of prolonged study of the mutual actions of the most seemingly diverse substances, and of minute and patient examination of the conditions under which nature performs her marvellous transmutations. The quest of the One Thing was fraught with peril, and was to be attempted only by those who had served a long and laborious apprenticeship.

In The Chemical Treatise of Thomas Norton, the Englishman, called Believe-me, or the Ordinal of Alchemy (15th century), the adept is warned not to disclose his secrets to ordinary people.

"You should carefully test and examine the life, character, and mental aptitudes of any person who would be initiated in this Art, and then you should bind him, by a sacred oath, not to let our Magistery be commonly or vulgarly known. Only when he begins to grow old and feeble, he may reveal it to one person, but not to more, and that one man must be virtuous.... If any wicked man should learn to practise the Art, the event would be fraught with great danger to Christendom. For such a man would overstep all bounds of moderation, and would remove from their hereditary thrones those legitimate princes who rule over the peoples of Christendom."

The results of the experimental examination of the compositions and properties of substances, made since the time of the alchemists, have led to the modern conception of the chemical element, and the isolation of about seventy or eighty different elements. No substance now called an element has been produced in the laboratory by uniting two, or more, distinct substances, nor has any been separated into two, or more, unlike portions. The only decided change which a chemical element has been caused to undergo is the combination of it with some other element or elements, or with a compound or compounds.

But it is possible that all the chemical elements may be combinations of different quantities of one primal element. Certain facts make this supposition tenable; and some chemists expect that the supposition will be proved to be correct. If the hypothetical primal element should be isolated, we should have fulfilled the aim of alchemy, and gained the One Thing; but the fulfilment would not be that whereof the alchemists dreamed.

Inasmuch as the alchemical Essence was thought of as the Universal Spirit to whose presence is due whatever degree of perfection any specific substance exhibits, it followed that the more perfect a substance the greater is the quantity of the Essence in it. But even in the most perfect substance found in nature—which substance, the alchemists said, is gold—the Essence is hidden by wrappings of specific properties which prevent the ordinary man from recognising it. Remove these wrappings from some special substance, and you have the perfect form of that thing; you have some portion of the Universal Spirit joined to the one general property of the class of things whereof the particular substance is a member. Then remove the class-property, often spoken of by the alchemists as the life, of the substance, and you have the Essence itself.

The alchemists thought that to every thing, or at any rate to every class of things, there corresponds a more perfect form than that which we see and handle; they spoke of gold, and the gold of the Sages; mercury, and the mercury of the Philosophers; sulphur, and the heavenly sulphur of him whose eyes are opened.

To remove the outer wrappings of ordinary properties which present themselves to the untrained senses, was regarded by the alchemists to be a difficult task; to tear away the soul (the class-property) of a substance, and yet retain the Essence which made that substance its dwelling place, was possible only after vast labour, and by the use of the proper agent working under the proper conditions. An exceedingly powerful, delicate, and refined agent was needed; and the mastery of the agent was to be acquired by bitter experience, and, probably, after many disappointments.

"Gold," an alchemist tells us, "does not easily give up its nature, and will fight for its life; but our agent is strong enough to overcome and kill it, and then it also has the power to restore it to life, and to change the lifeless remains into a new and pure body."

Thomas Norton, the author of The Ordinal of Alchemy, writing in the 15th century, says the worker in transmutations is often tempted to be in a hurry, or to despair, and he is often deceived. His servants will be either stupid and faithful, or quick-witted and false. He may be robbed of everything when his work is almost finished. The only remedies are infinite patience, a sense of virtue, and sound reason. "In the pursuit of our Art," he says, "you should take care, from time to time, to unbend your mind from its sterner employments with some convenient recreation."

The choice of workmen to aid in the mechanical parts of the quest was a great trouble to the alchemists. On this subject Norton says—"If you would be free from all fear over the gross work, follow my counsel, and never engage married men; for they soon give in and pretend they are tired out.... Hire your workmen for certain stipulated wages, and not for longer periods than twenty-four hours at a time. Give them higher wages than they would receive elsewhere, and be prompt and ready in your payments."

Many accounts are given by alchemical writers of the agent, and many names are bestowed on it. The author of A Brief Guide to the Celestial Ruby speaks thus of the agent—"It is our doorkeeper, our balm, our honey, oil, urine, maydew, mother, egg, secret furnace, oven, true fire, venomous dragon, Theriac, ardent wine, Green Lion, Bird of Hermes, Goose of Hermogenes, two-edged sword in the hand of the Cherub that guards the Tree of Life.... It is our true secret vessel, and the Garden of the Sages in which our sun rises and sets. It is our Royal Mineral, our triumphant vegetable Saturnia, and the magic rod of Hermes, by means of which he assumes any shape he likes."

Sometimes we are told that the agent is mercury, sometimes that it is gold, but not common mercury or common gold. "Supplement your common mercury with the inward fire which it needs, and you will soon get rid of all superfluous dross." "The agent is gold, as highly matured as natural and artificial digestion can make it, and a thousand times more perfect than the common metal of that name. Gold, thus exalted, radically penetrates, tinges, and fixes metals."

The alchemists generally likened the work to be performed by their agent to the killing of a living thing. They constantly use the allegory of death, followed by resurrection, in describing the steps whereby the Essence was to be obtained, and the processes whereby the baser metals were to be partially purified. They speak of the mortification of metals, the dissolution and putrefaction of substances, as preliminaries to the appearance of the true life of the things whose outward properties have been destroyed. For instance, Paracelsus says: "Destruction perfects that which is good; for the good cannot appear on account of that which conceals it." The same alchemist speaks of rusting as the mortification of metals; he says: "The mortification of metals is the removal of their bodily structure.... The mortification of woods is their being turned into charcoal or ashes."

Paracelsus distinguishes natural from artificial mortification, "Whatever nature consumes," he says, "man cannot restore. But whatever man destroys man can restore, and break again when restored." Things which had been mortified by man's device were considered by Paracelsus not to be really dead. He gives this extraordinary illustration of his meaning: "You see this is the case with lions, which are all born dead, and are first vitalised by the horrible noise of their parents, just as a sleeping person is awakened by a shout."

The mortification of metals is represented in alchemical books by various images and allegories. Fig. I. is reduced from a cut in a 16th century work, The Book of Lambspring, a noble ancient Philosopher, concerning the Philosophical Stone.

**Here the father devours the son;
The soul and spirit flow forth from the body.**

FIG. I.

The image used to set forth the mortification of metals is a king swallowing his son. Figs. II. and III. are reduced from Basil Valentine's Twelve Keys. Both of these figures represent the process of mortification by images connected with death and burial.

FIG. II.

In his explanation (?) of these figures, Basil Valentine says:—

"Neither human nor animal bodies can be multiplied or propagated without decomposition; the grain and all vegetable seed, when cast into the ground, must decay before it can spring up again; moreover, putrefaction imparts life to many worms and other animalculæ.... If bread is placed in honey, and suffered to decay, ants are generated ... maggots are also developed by the decay of nuts, apples, and pears. The same thing may be observed in regard to vegetable life. Nettles and other weeds spring up where no such seed has ever been sown. This occurs only by putrefaction. The reason is that the soil in such places is so disposed, and, as it were, impregnated, that it produces these fruits; which is a result of the properties of sidereal influences; consequently the seed is spiritually produced in the earth, and putrefies in the earth, and by the operation of the elements generates corporeal matter according to the species of nature. Thus the stars and the elements may generate new spiritual, and ultimately, new vegetable seed, by means of putrefaction.... Know that, in like manner, no metallic seed can develop, or multiply, unless the said seed, by itself alone, and without the introduction of any foreign substance, be reduced to a perfect putrefaction."

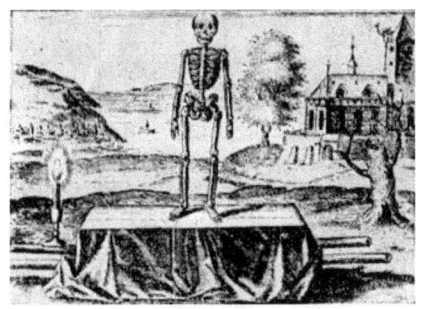

FIG. III.

The action of the mineral agent in perfecting substances is often likened by the alchemists to the conjoining of the male and the female, followed by the production of offspring. They insist on the need of a union of two things, in order to produce something more perfect than either. The agent, they say, must work upon something; alone it is nothing.

The methods whereby the agent is itself perfected, and the processes wherein the agent effects the perfecting of the less perfect things, were divided into stages by the alchemists. They generally spoke of these stages as Gates, and enumerated ten or sometimes twelve of them. As examples of the alchemical description of these gates, I give some extracts from A Brief Guide to the Celestial Ruby.

The first gate is Calcination, which is "the drying up of the humours"; by this process the substance "is concocted into a black powder which is yet unctuous, and retains its radical humour." When gold passes through this gate, "We observe in it two natures, the fixed and the volatile, which we liken to two serpents." The fixed nature is likened to a serpent without wings; the volatile, to a serpent with wings: calcination unites these two into one. The second gate, Dissolution, is likened to death and burial; but the true Essence will appear glorious and beautiful when this gate is passed. The worker is told not to be discouraged by this apparent death. The mercury of the sages is spoken of by this author as the queen, and gold as the king. The king dies for love of the queen, but he is revived by his spouse, who is made fruitful by him and brings forth "a most royal son."

Figs. IV. and V. are reduced from The Book of Lambspring; they express the need of the conjunction of two to produce one.

Here you behold a great marvel— Two Lions are joined into one.

The spirit and soul must be united in their body.
FIG. IV.

After dissolution came Conjunction, wherein the separated elements were combined. Then followed Putrefaction, necessary for the germination of the seed which had been produced by calcination, dissolution, and conjunction. Putrefaction was followed by Congelation and Citation. The passage through the next gate, called Sublimation, caused the body to become spiritual, and the spiritual to be made corporal. Fermentation followed, whereby the substance became soft and flowed like wax. Finally, by Exaltation, the Stone was perfected.

Here are two birds, great and strong—the body and spirit; one devours the other.

Let the body be placed in horse-dung, or a warm bath, the spirit having been extracted from it. The body has become white by the process, the spirit red by our art. All that exists tends towards perfection, and thus is the Philosopher's Stone prepared.
FIG. V.

The author of The Open Entrance speaks of the various stages in the perfecting of the agent as regimens. The beginning of the heating of gold with mercury is likened to the king stripping off his golden garments and descending into the fountain; this is the regimen of Mercury. As the heating is continued, all becomes black; this is the regimen of Saturn. Then is noticed a play of many colours; this is the regimen of Jupiter: if the heat is not regulated properly, "the young ones of the crow will go back to the nest." About the end of the fourth month you will see "the sign of the waxing moon," and all becomes white; this is the regimen of the Moon. The white colour gives place to purple and green; you are now in the regimen of Venus. After that, appear all the colours of the rainbow, or of a peacock's tail; this is the regimen of Mars. Finally the colour becomes orange and golden; this is the regimen of the Sun.

The reader may wish to have some description of the Essence. The alchemists could describe it only in contraries. It had a bodily form, but its method of working was spiritual. In The Sodic Hydrolith, or Water Stone of the Wise we are told:—

> "The stone is conceived below the earth, born in the earth, quickened in heaven, dies in time, and obtains eternal glory.... It is bluish-grey and green.... It flows like water, yet it makes no wet; it is of great weight, and is small."

Philalethes says, in A Brief Guide to the Celestial Ruby: "The Philosopher's Stone is a certain heavenly, spiritual, penetrative, and fixed substance, which brings all metals to the perfection of gold or silver (according to the quality of the Medicine), and that by natural methods, which yet in their effects transcend Nature.... Know then that it is called a stone, not because it is like a stone, but only because, by virtue of its fixed nature, it resists the action of fire as successfully as any stone. In species it is gold, more pure than the purest; it is fixed and incombustible like a stone, but its appearance is that of very fine powder, impalpable to the touch, sweet to the taste, fragrant to the smell, in potency a most penetrative spirit, apparently dry and yet unctuous, and easily capable of tinging a plate of metal.... If we say that its nature is spiritual, it would be no more than the truth; if we described it as corporeal, the expression would be equally correct."

The same author says: "There is a substance of a metalline species which looks so cloudy that the universe will have nothing to do with it. Its visible form is vile; it defiles metalline bodies, and no one can readily imagine that the pearly drink of bright Phœbus should spring from thence. Its components are a most pure and tender mercury, a dry incarcerate sulphur, which binds it and restrains fluxation.... Know this subject, it is the sure basis of all our secrets.... To deal plainly, it is the child of Saturn, of mean price and great venom.... It is not malleable, though metalline. Its colour is sable, with intermixed argent which mark the sable fields with veins of glittering argent."

In trying to attach definite meanings to the alchemical accounts of Principles, Elements, and the One Thing, and the directions which the alchemists give for changing one substance into others, we are very apt to be misled by the use of such an expression as the transmutation of the elements. To a chemist that phrase means the change of an element into another element, an element being a definite substance, which no one has been able to produce by the combination of two or more substances unlike itself, or to separate into two or more substances unlike itself. But whatever may have been the alchemical meaning of the word element, it was certainly not that given to the same word to-day. Nor did the word transmutation mean to the alchemist what it means to the chemist.

The facts which are known at present concerning the elements make unthinkable such a change as that of lead into silver; but new facts may be discovered which will make possible the separation of lead into things unlike itself, and the production of silver by the combination of some of these constituents of lead. The alchemist supposed he knew such facts as enabled him not only to form a mental picture of the change of lead into silver, or tin into gold, but also to assert that such changes must necessarily happen, and to accomplish them. Although we are quite sure that the alchemist's facts were only imaginings, we ought not to blame him for his reasoning on what he took to be facts.

Every metal is now said to be an element, in the modern meaning of that word: the alchemist regarded the metals as composite substances; but he also thought of them as more simple than many other things. Hence, if he was able to transmute one metal into another, he would have strong evidence in support of his general conception of the unity of all things. And, as transmutation meant, to the alchemist, the bringing of a substance to the condition of greatest perfection possible for that substance, his view of the unity of nature might be said to be proved if he succeeded in changing one of the metals, one of these comparatively simple substances, into the most perfect of all metals, that is, into gold.

The transmutation of the baser metals into gold thus came to be the practical test of the justness of the alchemical scheme of things.

Some alchemists assert they had themselves performed the great transmutation; others tell of people who had accomplished the work. The following story is an example of the accounts given of the making of gold. It is taken from John Frederick Helvetius' Golden Calf, which the world worships and adores (17th century):—

> "On the 27th December 1666, in the forenoon, there came to my house a certain man, who was a complete stranger to me, but of an honest grave countenance, and

an authoritative mien, clothed in a simple garb.... He was of middle height, his face was long and slightly pock-marked, his hair was black and straight, his chin close-shaven, his age about forty-three or forty-four, and his native province, as far as I could make out, North Holland. After we had exchanged salutations, he asked me whether he might have some conversation with me. He wished to say something to me about the Pyrotechnic Art, as he had read one of my tracts (directed against the Sympathetic Powder of Dr Digby), in which I hinted a suspicion whether the Grand Arcanum of the Sages was not after all a gigantic hoax. He, therefore, took that opportunity of asking me whether I could not believe that such a grand mystery might exist in the nature of things, by means of which a physician could restore any patient whose vitals were not irreparably destroyed. I answered, 'Such a medicine would be a most desirable acquisition for any physician; nor can any man tell how many secrets there may be hidden in Nature; yet, though I have read much about the truth of this art, it has never been my good fortune to meet with a real master of the alchemical science.' ... After some further conversation, the Artist Elias (for it was he) thus addressed me: 'Since you have read so much in the works of the alchemists about this stone, its substance, its colour and its wonderful effects, may I be allowed the question, whether you have not prepared it yourself?' On my answering his question in the negative, he took out of his bag a cunningly-worked ivory box, in which were three large pieces of substance resembling glass, or pale sulphur, and informed me that here was enough of the tincture for the production of twenty tons of gold. When I had held the precious treasure in my hand for a quarter of an hour (during which time I listened to a recital of its wonderful curative properties), I was compelled to restore it to its owner, which I could not help doing with a certain degree of reluctance.... My request that he would give me a piece of his stone (though it were no larger than a coriander seed), he somewhat brusquely refused, adding, in a milder tone, that he could not give it me for all the wealth I possessed, and that not on account of its great preciousness, but for some other reason which it was not lawful for him to divulge.... Then he inquired whether I could not show him into a room at the back of the house, where we should be less liable to the observation of passers-by. On my conducting him into the state parlour (which he entered without wiping his dirty boots), he demanded of me a gold coin, and while I was looking for it, he produced from his breast pocket a green silk handkerchief, in which were folded up five medals, the gold of which was infinitely superior to that of my gold piece." Here follows the inscriptions on the medals. "I was filled with admiration, and asked my visitor whence he had obtained that wonderful knowledge of the whole world. He replied that it was a gift freely bestowed on him by a friend who had stayed a few days at his house." Here follows the stranger's account of this friend's experiments. "When my strange visitor had concluded his narrative, I besought him to give me a proof of his assertion, by performing the transmutatory operation on some metals in my presence. He answered evasively, that he could not do so then, but that he would return in three weeks, and that, if he was then at liberty to do so, he would show me something that would make me open my eyes. He appeared punctually to the promised day, and invited me to take a walk with him, in the course of which we discoursed profoundly on the secrets of Nature in fire, though I noticed that my companion was very chary in imparting information about the Grand Arcanum.... At last I asked him point blank to show me the transmutation of metals. I besought him to come and dine with me, and to spend the night at my house; I entreated; I expostulated; but in vain. He remained firm. I reminded him of his promise. He retorted that his promise had been conditional upon his being permitted to reveal the secret to me. At last, however, I prevailed upon him to give me a piece of his precious stone—a piece no larger than a grain of rape seed.... He bid me take half an ounce of lead ... and melt it in the crucible; for the Medicine would certainly not tinge more of the base metal than it was sufficient for.... He promised to return at nine o'clock the next morning.... But at the stated hour on the following day he did not make his appearance; in his stead, however, there came, a few hours later, a stranger, who told me that his friend the artist was unavoidably detained, but that he would call at three o'clock in the afternoon. The afternoon came; I waited for him till half-past seven o'clock. He did not appear. Thereupon my wife came and tempted me to try the transmutation myself. I determined however to wait till the morrow. On the morrow ... I asked my wife to put the tincture in wax, and I myself ... prepared six drachms of lead; I then cast the tincture, enveloped as it was in wax, on the lead; as soon as it was melted, there was a hissing sound and a slight effervescence, and after a quarter of an hour I found that the whole mass of lead had been turned into the finest gold.... We immediately took it to the goldsmith, who at once declared it the finest gold he had ever seen, and offered to pay fifty florins an ounce for it." He then describes various tests which were made to prove the purity of the gold. "Thus I have unfolded to you the whole story from beginning to end. The gold I still retain in my possession, but I cannot tell you what has become of the Artist Elias."

CHAPTER VI.

ALCHEMY AS AN EXPERIMENTAL ART.

A modern writer, Mr A.E. Waite, in his Lives of the Alchemystical Philosophers, says: "The physical theory of transmutation is based on the composite character of the metals, on their generation in the bowels of the earth, and on the existence in nature of a pure and penetrating matter which applied to any substance exalts and perfects it after its own kind." It must he admitted that the alchemists could cite many instances of transmutations which seemed to lead to the conclusion, that there is no difference of kind between the metals and other substances such as water, acids, oils, resins, and wood. We are able to-day to effect a vast number of transformations wherein one substance is exchanged for another, or made to take the place of another. We can give fairly satisfactory descriptions of these changes; and, by comparing them one with another, we are able to express their essential features in general terms which can be applied to each particular instance. The alchemists had no searching knowledge of what may be called the mechanism of such changes; they gave an explanation of them which we must call incorrect, in the present state of our knowledge. But, as Hoefer says in his Histoire de la Chimie, "to jeer at [the alchemical] theory is to commit at once an anachronism and an injustice.... Unless the world should finish to-morrow, no one can have the pretension to suppose that our contemporaries have said the last word of science, and nothing will remain for our descendants to discover, no errors for them to correct, no theories for them to set straight."

FIG. VI. See p. 90.

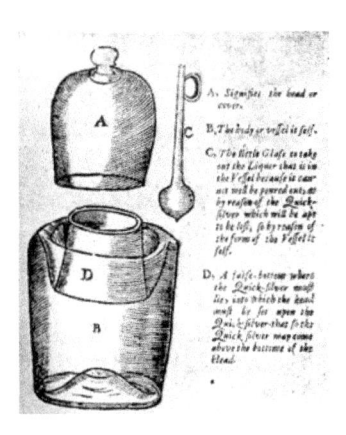

FIG. VII. See p. 90.

FIG. VIII. See p. 91.

What kind of experimental evidence could an alchemist furnish in support of his theory of transmutation? In answering this question, I cannot do better than give a condensed rendering of certain pages in Hoefer's Histoire de la Chimie.

The reader is supposed to be present at experiments conducted in the laboratory of a Grand Master of the Sacred Art in the 5th or 6th century.

Experiment.—Ordinary water is boiled in an open vessel; the water is changed to a vapour which disappears, and a white powdery earth remains in the vessel.

Conclusion.—Water is changed into air and earth.

Did we not know that ordinary water holds certain substances in solution, and that boiling water acts on the vessel wherein it is boiled, we should have no objection to urge against this conclusion.

It only remained to transmute fire that the transmutation of the four elements might be completed.

Experiment.—A piece of red-hot iron is placed in a bell-jar, filled with water, held over a basin containing water; the volume of the water decreases, and the air in the bell-jar takes fire when a lighted taper is brought into it.

Conclusion.—Water is changed into fire.

That interpretation was perfectly reasonable at a time when the fact was unknown that water is composed of two gaseous substances; that one of these (oxygen) is absorbed by the iron, and the other (hydrogen) collects in the bell-jar, and ignites when brought into contact with a flame.

Experiment.—Lead, or any other metal except gold or silver, is calcined in the air; the metal loses its characteristic properties, and is changed into a powdery substance, a kind of cinder or calx. When this cinder, which was said to be the result of the death of the metal, is heated in a crucible with some grains of wheat, one sees the metal revive, and resume its original form and properties.

Conclusion.—The metal which had been destroyed is revivified by the grains of wheat and the action of fire.

Is this not to perform the miracle of the resurrection?

No objection can he raised to this interpretation, as long as we are ignorant of the phenomena of oxidation, and the reduction of oxides by means of carbon, or organic substances rich in carbon, such as sugar, flour, seeds, etc. Grains of wheat were the symbol of life, and, by extension, of the resurrection and eternal life.

FIG. IX. See p. 91.

Experiment.—Ordinary lead is calcined in a cupel made of cinders or powdered bones; the lead is changed to a cinder which disappears into the cupel, and a button of silver remains.

Conclusion.—The lead has vanished; what more natural than the conclusion that it has been transformed into silver? It was not known then that all specimens of lead contain more or less silver.

FIG. X. See p. 92.

Experiment.-The vapour of arsenic bleaches copper. This fact gave rise to many allegories and enigmas concerning the means of transforming copper into silver.

Sulphur, which acts on metals and changes many of them into black substances, was looked on as a very mysterious thing. It was with sulphur that the coagulation (solidification) of mercury was effected.

Experiment.—Mercury is allowed to fall, in a fine rain, on to melted sulphur; a black substance is produced; this black substance is heated in a closed vessel, it is volatilised and transformed into a beautiful red solid.

One could scarcely suppose that the black and the red substances are identical, if one did not know that they are composed of the same quantities of the same elements, sulphur and mercury.

How greatly must this phenomenon have affected the imagination of the chemists of ancient times, always so ready to be affected by everything that seemed supernatural!

Black and red were the symbols of darkness and light, of the evil and the good principle; and the union of these two principles represented the moral order. At a later time the idea helped to establish the alchemical doctrine that sulphur and mercury are the Principles of all things.

Experiment.—Various organic substances are analysed by heating in a distillation-apparatus; the products are, in each case, a solid residue, liquids which distil off, and certain spirits which are disengaged.

The results supported the ancient theory which asserted that earth, water, air, and fire are the four Elements of the world. The solid residue represented earth; the liquid products of the distillation, water; and the spirituous substances, air. Fire was regarded sometimes as the means of purification, sometimes as the soul, or invisible part, of all substances.

Experiment.-A strong acid is poured on to copper. The metal is attacked, and at last disappears, giving place to a green liquid, as transparent as water. A thin sheet of iron is plunged into the liquid; the copper re-appears, and the iron vanishes.

What more simple than to conclude that the iron has been transformed into copper?

Had lead, silver, or gold been used in place of copper, one would have said that the iron was transformed into lead, silver, or gold.

In their search for "the pure and penetrating matter which applied to any substance exalts and perfects it after its own kind," the alchemists necessarily made many inventions, laid the foundation of many arts and manufactures, and discovered many facts of importance in the science of chemistry.

The practitioners of the Sacred Art of Egypt must have been acquainted with many operations which we now class as belonging to applied chemistry; witness, their jewellery, pottery, dyes and pigments, bleaching, glass-making, working in metals and alloys, and their use of spices, essential oils, and soda in embalming, and for other purposes.

During the centuries when alchemy flourished, gunpowder was invented, the art of printing was established, the compass was brought into use, the art of painting and staining glass was begun and carried to perfection, paper was made from rags, practical metallurgy advanced by leaps and bounds, many new alloys of metals came into use, glass mirrors were manufactured, and considerable advances were made in practical medicine and sanitation.

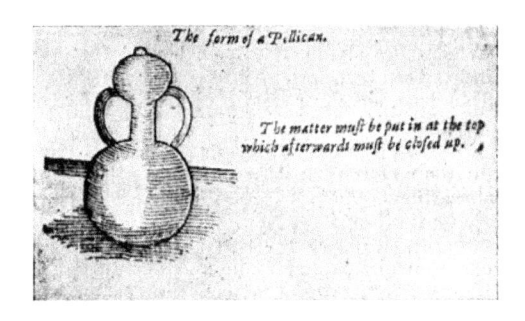

FIG. XI. See p. 92.

Basil Valentine, who was one of the greatest alchemists of the 16th century, discovered many of the properties of the metal antimony, and prepared and examined many compounds of that metal; he made green vitriol from pyrites, brandy from fermented grape-juice, fulminating gold, sulphide of potash, and spirits of salt; he made and used baths of artificial mineral waters, and he prepared various metals by what are now called wet methods, for instance, copper, by immersing plates of iron in solutions of bluestone. He examined the air of mines, and suggested practical methods for determining whether the air in a mine was respirable. Hoefer draws attention to a remarkable observation recorded by this alchemist. Speaking of the "spirit of mercury," Basil Valentine says it is "the origin of all the metals; that spirit is nothing else than an air flying here and there without wings; it is a moving wind, which, after it has been chased from its home of Vulcan (that is, fire), returns to the chaos; then it expands and passes into the region of the air from whence it had come." As Hoefer remarks, this is perhaps one of the earliest accounts of the gas discovered by Priestley and studied by Lavoisier, the gas we now call oxygen, and recognise as of paramount importance in chemical reactions.

FIG. XII. See p. 92.

Besides discovering and recording many facts which have become part and parcel of the science of chemistry, the alchemists invented and used various pieces of apparatus, and conducted many operations, which are still employed in chemical laboratories. I shall reproduce illustrations of some of these processes and pieces of apparatus, and quote a few of the directions, given in a book, published in 1664, called The Art of Distillation, by John French, Dr. in Physick.

The method recommended by French for hermetically sealing the neck of a glass vessel is shown in Fig. VI. p. 80. The neck of the vessel is surrounded by a tray containing burning coals; when the glass melts it is cut off by shears, and then closed by tongs, which are made hot before use.

Fig. VII. p. 81, represents a method for covering an open vessel, air-tight, with a receptacle into which a substance may be sublimed from the lower vessel. The lettering explains the method of using the apparatus.

French gives very practical directions and much sound advice for conducting distillations of various kinds. The following are specimens of his directions and advice:—

> "When you put water into a seething Balneum wherein there are glasses let it be hot, or else thou wilt endanger the breaking of the glasses.

> "When thou takest any earthen, or glass vessel from the fire, expose it not to the cold aire too suddenly for fear it should break.

> "In all your operations diligently observe the processes which you read, and vary not a little from them, for sometimes a small mistake or neglect spoils the whole operation, and frustrates your expectations.

> "Try not at first experiments of great cost, or great difficulty; for it will be a great discouragement to thee, and thou wilt be very apt to mistake.

> "If any one would enter upon the practices of Chymistry, let him apply himself to some expert artist for to be instructed in the manual operation of things; for by this means he will learn more in two months, than he can by his practice and study in seven years, as also avoid much pains and cost, and redeem much time which else of necessity he will lose."

Fig. VIII. p. 82, represents a common cold still, and Fig. IX. p. 84, is a sketch of an apparatus for distilling by the aid of boiling water. The bath wherein the vessels are placed in Fig. IX. was called by the alchemists balneum Mariae, from Mary the Jewess, who is mentioned in the older alchemical writings, and is supposed to have invented an apparatus of this character. Nothing definite is known of Mary the Jewess. A writer of the 7th century says she was initiated in the sacred art in the temple of Memphis; a legend prevailed among some of the alchemists that she was the sister of Moses.

Fig. X. p. 85, represents methods of distilling with an apparatus for cooling the volatile products; the lower vessel is an alembic, with a long neck, the upper part of which passes through a vessel containing cold water.

Fig XIII. See p. 94.

Fig. XI. p. 88, shows a pelican, that is a vessel wherein a liquid might be heated for a long time, and the volatile products be constantly returned to the original vessel.

Fig. XII. p. 89, represents a retort with a receiver.

Some of the pieces of apparatus for distilling, which are described by French, are shown in the following figures. Besides describing apparatus for distilling, subliming, and other processes in the laboratory, French gives directions for making tinctures, essences, essential oils, spirits of salt, and pure saltpetre, oil of vitriol, butter of antimony, calces (or as we now say, oxides) of metals, and many other substances. He describes processes for making fresh water from salt, artificial mineral water, medicated hot baths for invalids (one of the figures represents an apparatus very like those advertised to-day as "Turkish baths at home"), and artificial precious stones; he tells how to test minerals, and make alloys, and describes the preparation of many substances made from gold and silver. He also gives many curious receipts; for instance, "To make Firre-trees appear in Turpentine," "To make a Plant grow in two or three hours," "To make the representation of the whole world in a Glass," "To extract a white Milkie substance from the raies of the Moon."

FIG. XIV. See p. 94.

The process of making oil of vitriol, by burning sulphur under a hood fitted with a side tube for the outflow of the oil of vitriol, is represented in Fig. XIII. p. 92.

Fig. XIV. p. 93, is interesting; it is an apparatus for rectifying spirits, by distilling, and liquefying only the most volatile portions of the distillate. The spirituous liquor was heated, and the vapours caused to traverse a long zigzag tube, wherein the less volatile portions condensed to liquid, which flowed back into the vessel; the vapour then passed into another vessel, and then through a second zigzag tube, and was finally cooled by water, and the condensed liquid collected. This apparatus was the forerunner of that used to-day, for effecting the separation of liquids which boil at different temperatures, by the process called fractional distillation.

We should never forget that the alchemists were patient and laborious workers, their theories were vitally connected with their practice, and there was a constant action and reaction between their general scheme of things and many branches of what we now call chemical manufactures. We may laugh at many of their theories, and regret that much useless material was accumulated by them; we may agree with Boyle (end of 17th century) when he likens the "hermetick philosophers," in their search for truth, to "the navigators of Solomon's Tarshish fleet, who brought home from their long and tedious voyages, not only gold, and silver, and ivory, but apes and peacocks too; for so the writings of several of your hermetick philosophers present us, together with divers substantial and noble experiments, theories, which either like peacocks' feathers make a great show but are neither solid nor useful; or else like apes, if they have some appearance of being rational, are blemished with some absurdity or other, that, when they are attentively considered make them appear ridiculous." But however we may condemn their method, because it rested on their own conception of what the order of nature must be, we cannot but praise their assiduity in conducting experiments and gathering facts.

As Bacon says, in De Augmentis Scientiarum:

> "Alchemy may be compared to the man who told his sons that he had left them gold buried somewhere in his vineyard; where they by digging found no gold, but by turning up the mould about the roots of the vines, procured a plentiful vintage. So the search and endeavours to make gold have brought many useful inventions and instructive experiments to light."

CHAPTER VII.

THE LANGUAGE OF ALCHEMY

The vagueness of the general conceptions of alchemy, and the attribution of ethical qualities to material things by the alchemists, necessarily led to the employment of a language which is inexact, undescriptive, and unsuggestive to modern ears. The same name was given to different things, and the same thing went under many names. In Chapter IV. I endeavoured to analyse two terms which were constantly used by the alchemists to convey ideas of great importance, the terms Element and Principle. That attempt sufficed, at any rate, to show the vagueness of the ideas which these terms were intended to express, and to make evident the inconsistencies between the meanings given to the words by different alchemical writers. The story quoted in Chapter III., from Michael Sendivogius, illustrates the difficulty which the alchemists themselves had in understanding what they meant by the term Mercury; yet there is perhaps no word more often used by them than that. Some of them evidently took it to mean the substance then, and now, called mercury; the results of this literal interpretation were disastrous; others thought of mercury as a substance which could be obtained, or, at any rate, might be obtained, by repeatedly distilling ordinary mercury, both alone and when mixed with other substances; others used the word to mean a hypothetical something which was liquid but did not wet things, limpid yet capable of becoming solid, volatile yet able to prevent the volatilisation of other things, and white, yet ready to cause other white things to change their colour; they thought of this something, this soul of mercury, as having properties without itself being tangible, as at once a substance and not a substance, at once a bodily spirit and a spiritual body.

It was impossible to express the alchemical ideas in any language save that of far-fetched allegory. The alchemical writings abound in such allegories. Here are two of them.

The first allegory is taken from The Twelve Keys, of Basilius Valentinus, the Benedictine:—

> "The eleventh key to the knowledge of the augmentation of our Stone I will put before you in the form of a parable.
>
> "There lived in the East a gilded knight, named Orpheus, who was possessed of immense wealth, and had everything that heart can wish. He had taken to wife his own sister, Euridice, who did not, however, bear him any

children. This he regarded as the punishment of his sin in having wedded his own sister, and was instant in prayer to God both by day and by night, that the curse might be taken from him. One night when he was buried in a deep sleep, there came to him a certain winged messenger, named Phœbus, who touched his feet, which were very hot, and said: 'Thou noble knight, since thou hast wandered through many cities and kingdoms and suffered many things at sea, in battle, and in the lists, the heavenly Father has bidden me make known to thee the following means of obtaining thy prayer: Take blood from thy right side, and from the left side of thy spouse. For this blood is the heart's blood of your parents, and though it may seem to be of two kinds, yet, in reality, it is only one. Mix the two kinds of blood, and keep the mixture tightly enclosed in the globe of the seven wise Masters. Then that which is generated will be nourished with its own flesh and blood, and will complete its course of development when the Moon has changed for the eighth time. If thou repeat this process again and again, thou shalt see children's children, and the offspring of thy body shall fill the world.' When Phœbus had thus spoken, he winged his flight heavenward. In the morning the knight arose and did the bidding of the celestial messenger, and God gave to him and to his wife many children, who inherited their father's glory, wealth, and knightly honours from generation to generation."

In the "Dedicatory Epistle" to his Triumphal Chariot of Antimony, Basil Valentine addresses his brother alchemists as follows:—

"Mercury appeared to me in a dream, and brought me back from my devious courses to the one way. 'Behold me clad not in the garb of the vulgar, but in the philosopher's mantle.' So he said, and straightway began to leap along the road in headlong bounds. Then, when he was tired, he sat down, and, turning to me, who had followed him in the spirit, bade me mark that he no longer possessed that youthful vigour with which he would at the first have overcome every obstacle, if he had not been allowed a free course. Encouraged by his friendly salutation, I addressed him in the following terms: 'Mercury, eloquent scion of Atlas, and father of all Alchemists, since thou hast guided me hitherto, shew me, I pray thee, the way to those

Blessed Isles, which thou hast promised to reveal to all thine elect children. 'Dost thou remember,' he replied, that when I quitted thy laboratory, I left behind me a garment so thoroughly saturated with my own blood, that neither the wind could efface it, nor all-devouring time destroy its indelible essence? Fetch it hither to me, that I may not catch a chill from the state of perspiration in which I now am; but let me clothe myself warmly in it, and be closely incited thereto, so that I may safely reach my bride, who is sick with love. She has meekly borne many wrongs, being driven through water and fire, and compelled to ascend and descend times without number—yet has she been carried through it all by the hope of entering with me the bridal chamber, wherein we expect to beget a son adorned from his birth with the royal crown which he may not share with others. Yet may he bring his friends to the palace, where sits enthroned the King of Kings, who communicates his dignity readily and liberally to all that approach him.'

"I brought him the garment, and it fitted him so closely, that it looked like an iron skin securing him against all the assaults of Vulcan. 'Let us proceed,' he then said, and straightway sped across the open field, while I boldly strove to keep up with my guide.

"Thus we reached his bride, whose virtue and constancy were equal to his own. There I beheld their marvellous conjugal union and nuptial consummation, whence was born the son crowned with the royal diadem. When I was about to salute him as King of Kings and Lord of Lords, my Genius stood by me and warned me not to be deceived, since this was only the King's forerunner, but not the King himself whom I sought.

"When I heard the admonition, I did not know whether to be sad or joyful. 'Depart,' then said Mercury, 'with this bridal gift, and when you come to those disciples who have seen the Lord himself, show them this sign.' And therewith he gave me a gold ring from his son's finger. 'They know the golden branch which must be consecrated to Proserpina before you can enter the palace of Pluto. When he sees this ring, perhaps one will open to you with a word the door of that chamber, where sits enthroned in

> his magnificence the Desire of all Nations, who is known only to the Sages.'
>
> "When he had thus spoken, the vision vanished, but the bridal gift which I still held in my hand shewed me that it had not been a mere dream. It was of gold, but to me more precious than the most prized of all metals. Unto you I will shew it when I am permitted to see your faces, and to converse with you freely. Till that earnestly wished-for time, I bid you farewell."

One result of the alchemical modes of expression was, that he who tried to follow the directions given in alchemical books got into dire confusion. He did not know what substances to use in his operations; for when he was told to employ "the homogeneous water of gold," for example, the expression might mean anything, and in despair he distilled, and calcined, and cohobated, and tried to decompose everything he could lay hands on. Those who pretended to know abused and vilified those who differed from them.

In A Demonstration of Nature, by John A. Mehung (17th century), Nature addresses the alchemical worker in the following words:—

> "You break vials, and consume coals, only to soften your brains still more with the vapours. You also digest alum, salt, orpiment, and altrament; you melt metals, build small and large furnaces, and use many vessels; nevertheless I am sick of your folly, and you suffocate me with your sulphurous smoke.... You would do better to mind your own business, than to dissolve and distil so many absurd substances, and then to pass them through alembics, cucurbits, stills, and pelicans."

Henry Madathanas, writing in 1622, says:—

> "Then I understood that their purgations, sublimations, cementations, distillations, rectifications, circulations, putrefactions, conjunctions, calcinations, incinerations, mortifications, revivifications, as also their tripods, athanors, reverberatory alembics, excrements of horses, ashes, sand, stills, pelican-viols, retorts, fixations, etc., are mere plausible impostures and frauds."

The author of The Only Way (1677) says:

> "Surely every true Artist must look on this elaborate tissue of baseless operations as the merest folly, and can only

wonder that the eyes of those silly dupes are not at last opened, that they may see something besides such absurd sophisms, and read something besides those stupid and deceitful books.... I can speak from bitter experience, for I, too, toiled for many years ... and endeavoured to reach the coveted goal by sublimation, distillation, calcination, circulation, and so forth, and to fashion the Stone out of substances such as urine, salt, atrament, alum, etc. I have tried hard to evolve it out of hairs, wine, eggs, bones, and all manner of herbs; out of arsenic, mercury, and sulphur, and all the minerals and metals.... I have spent nights and days in dissolving, coagulating, amalgamating, and precipitating. Yet from all these things I derived neither profit nor joy."

Another writer speaks of many would-be alchemists as "floundering about in a sea of specious book-learning."

If alchemists could speak of their own processes and materials as those authors spoke whom I have quoted, we must expect that the alchemical language would appear mere jargon to the uninitiated. In Ben Jonson's play The Alchemist, Surley, who is the sceptic of the piece, says to Subtle, who is the alchemist—

... Alchemy is a pretty kind of game, Somewhat like tricks o' the cards, to cheat a man With charming ... What else are all your terms, Whereon no one of your writers 'grees with other? Of your elixir, your lac virginis, Your stone, your med'cine, and your chrysosperme, Your sal, your sulphur, and your mercury, Your oil of height, your tree of life, your blood, Your marchesite, your tutie, your magnesia, Your toad, your crow, your dragon, and your panther; Your sun, your moon, your firmament, your adrop, Your lato, azoch, zernich, chibrit, heutarit, And then your red man, and your white woman, With all your broths, your menstrues, and materials, Of lye and egg-shells, women's terms, man's blood, Hair o' the head, burnt clout, chalk, merds, and clay, Powder of bones, scalings of iron, glass, And moulds of other strange ingredients, Would burst a man to name?

To which Subtle answers,

And all these named Intending but one thing; which art our writers Used to obscure their art. Was not all the knowledge Of the Egyptians writ in mystic symbols? Speak not the Scriptures oft in parables? Are not the choicest fables of the poets, That were the fountains and first springs of wisdom, Wrapp'd in perplexed allegories?

The alchemists were very fond of using the names of animals as symbols of certain mineral substances, and of representing operations in the laboratory by what may be called animal allegories. The yellow lion was the alchemical symbol of yellow sulphides, the red lion was synonymous with cinnabar, and the green lion meant salts of iron and of copper. Black sulphides were called eagles, and sometimes crows. When black sulphide of mercury is strongly heated, a red sublimate is obtained, which has the same composition as the black compound; if the temperature is not kept very high, but little of the red sulphide is produced; the alchemists directed to urge the fire, "else the black crows will go back to the nest."

The salamander was called the king of animals, because it was supposed that he lived and delighted in fire; keeping a strong fire alight under a salamander was sometimes compared to the purification of gold by heating it.

Fig. XV., reduced from The Book of Lambspring represents this process.

A salamander lives in the fire, which imparts to it a most glorious hue.

This is the reiteration, gradation, and amelioration of the Tincture, or Philosopher's Stone; and the whole is called its Augmentation.
FIG. XV.

The alchemists employed many signs, or shorthand expressions, in place of writing the names of substances. The following are a few of the signs which were used frequently.

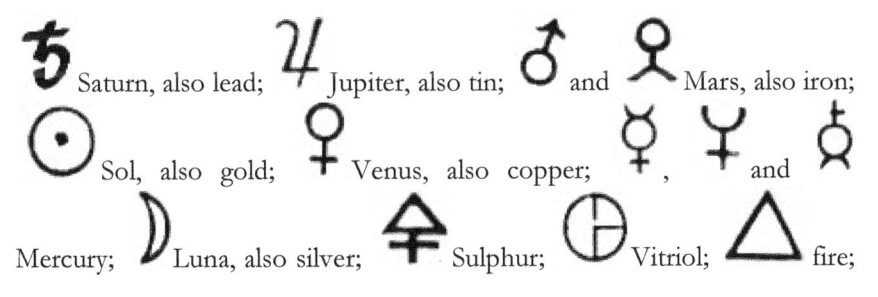

♄ Saturn, also lead; ♃ Jupiter, also tin; ♂ and ♀ Mars, also iron; ☉ Sol, also gold; ♀ Venus, also copper; ☿ , ☿ and ☿ Mercury; ☽ Luna, also silver; 🜍 Sulphur; 🜖 Vitriol; △ fire;

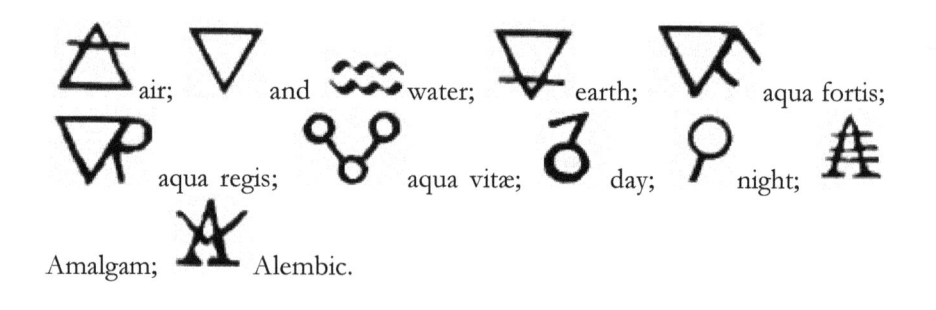

△ air; ▽ and 〰 water; ▽ earth; ⟁ aqua fortis; ⟁ aqua regis; ⚯ aqua vitæ; ☉ day; ♀ night; ♓ Amalgam; ⚚ Alembic.

CHAPTER VIII.

THE DEGENERACY OF ALCHEMY.

I have tried to show that alchemy aimed at giving experimental proof of a certain theory of the whole system of nature, including humanity. The practical culmination of the alchemical quest presented a threefold aspect; the alchemists sought the stone of wisdom, for by gaining that they gained the control of wealth; they sought the universal panacea, for that would give them the power of enjoying wealth and life; they sought the soul of the world, for thereby they could hold communion with spiritual existences, and enjoy the fruition of spiritual life.

The object of their search was to satisfy their material needs, their intellectual capacities, and their spiritual yearnings. The alchemists of the nobler sort always made the first of these objects subsidiary to the other two; they gave as their reason for desiring to make gold, the hope that gold might become so common that it would cease to be sought after by mankind. The author of An Open Substance says: "Would to God ... all men might become adepts in our art, for then gold, the common idol of mankind, would lose its value, and we should prize it only for its scientific teaching."

But the desire to make gold must always have been a very powerful incentive in determining men to attempt the laborious discipline of alchemy; and with them, as with all men, the love of money was the root of much evil. When a man became a student of alchemy merely for the purpose of making gold, and failed to make it—as he always did—it was very easy for him to pretend he had succeeded in order that he might really make gold by cheating other people. Such a man rapidly degenerated into a charlatan; he used the language of alchemy to cover his frauds, and with the hope of deluding his dupes by high-sounding phrases. And, it must be admitted, alchemy lent itself admirably to imposture. It promised unlimited wealth; it encouraged the wildest dreams of the seeker after pleasure; and over these dreams it cast the glamour of great ideas, the idea of the unity of nature, and the idea of communion with other spheres of life, of calling in the help of 'inheritors of unfulfilled renown,' and so it seemed to touch to fine issues the sordidness of unblushing avarice.

Moreover, the working with strange ingredients and odd-fashioned instruments, and the employment of mouth-filling phrases, and scraps of occult learning which seemed to imply unutterable things, gave just that pleasing dash of would-be wickedness to the process of consulting the alchemist which acts as a fascination to many people. The earnest person felt that by using the skill and knowledge of the alchemists, for what he deemed a good purpose, he was compelling the powers of evil to work for him and his objects.

It was impossible that such a system as alchemy should appear to the plain man of the middle ages, when the whole scheme of life and the universe rested on a magical basis, to be more than a kind of magic which hovered between the black magic of the Sorcerer and the white magic of the Church. Nor is it to be wondered at that a system which lends itself to imposture so easily as alchemy did, should be thought of by the plain man of modern times as having been nothing but a machinery of fraud.

It is evident from the Canon's Yeoman's Tale in Chaucer, that many of those who professed to turn the base metals into gold were held in bad repute as early as the 14th century. The "false chanoun" persuaded the priest, who was his dupe, to send his servant for quicksilver, which he promised to make into "as good silver and as fyn, As ther is any in youre purse or myn"; he then gave the priest a "crosselet," and bid him put it on the fire, and blow the coals. While the priest was busy with the fire,

This false chanoun—the foulè feend hym fecche!— Out of his bosom took a bechen cole, In which ful subtilly was maad an hole, And therinne put was of silver lemaille An ounce, and stoppéd was withouten faille The hole with wex, to kepe the lemaille in.

The "false chanoun" pretended to be sorry for the priest, who was so busily blowing the fire:—

Ye been right hoot, I se wel how ye swete; Have heer a clooth, and wipe awey the we't. And whylès that the preest wipèd his face, This chanoun took his cole with hardè grace, And leyde it above, upon the middèward Of the crosselet, and blew wel afterward. Til that the colès gonnè fastè brenne.

As the coal burned the silver fell into the "crosselet." Then the canon said they would both go together and fetch chalk, and a pail of water, for he would pour out the silver he had made in the form of an ingot. They locked the door, and took the key with them. On returning, the canon formed the chalk into a mould, and poured the contents of the crucible into it. Then he bade the priest,

Look what ther is, put in thin hand and grope, Thow fyndè shalt ther silver, as I hope. What, devel of hellè! Sholde it ellis be? Shavyng of silver silver is, parde! He putte his hand in, and took up a teyne Of silver fyn, and glad in every veyne Was this preest, when he saugh that it was so.

The conclusion of the Canon's Yeoman's Tale shows that, in the 14th century, there was a general belief in the possibility of finding the philosopher's stone, and effecting the transmutation, although the common practitioners of the art were regarded as deceivers. A disciple of Plato is supposed to ask his master to tell him the "namè of the privee stoon." Plato gives him certain directions, and tells him he must use magnasia; the disciple asks—

'What is Magnasia, good sire, I yow preye?' 'It is a water that is maad, I seye, Of elementés fourè,' quod Plato. 'Telle me the rootè, good sire,' quod he tho, Of that water, if it be yourè wille.' 'Nay, nay,' quod Plato, 'certein that I nylle; The philosophres sworn were everychoon That they sholden discovers it unto noon, Ne in no book it write in no manere, For unto Crist it is so lief and deere, That he wol nat that it discovered bee, But where it liketh to his deitee Man for tenspire, and eek for to deffende Whom that hym liketh; lo, this is the ende.'

The belief in the possibility of alchemy seems to have been general sometime before Chaucer wrote; but that belief was accompanied by the conviction that alchemy was an impious pursuit, because the transmutation of baser metals into gold was regarded as trenching on the prerogative of the Creator, to whom alone this power rightfully belonged. In his Inferno (which was probably written about the year 1300), Dante places the alchemists in the eighth circle of hell, not apparently because they were fraudulent impostors, but because, as one of them says, "I aped creative nature by my subtle art."

In later times, some of those who pretended to have the secret and to perform great wonders by the use of it, became rich and celebrated, and were much sought after. The most distinguished of these pseudo-alchemists was he who passed under the name of Cagliostro. His life bears witness to the eagerness of human beings to be deceived.

Joseph Balsamo was born in 1743 at Palermo, where his parents were tradespeople in a good way of business.[5] In the memoir of himself, which he wrote in prison, Balsamo seeks to surround his birth and parentage with mystery; he says, "I am ignorant, not only of my birthplace, but even of the parents who bore me.... My earliest infancy was passed in the town of Medina, in Arabia, where I was brought up under the name of Acharat."

When he was thirteen years of age, Balsamo's parents determined he should be trained for the priesthood, but he ran away from his school. He was then confined in a Benedictine monastery. He showed a remarkable taste for natural history, and acquired considerable knowledge of the use of drugs; but he soon tired of the discipline and escaped. For some years he wandered about in different parts of Italy, living by his wits and by cheating. A goldsmith consulted him about a hidden treasure; he pretended to invoke the aid of spirits, frightened the goldsmith, got sixty ounces of gold from him to carry on his incantations, left him in the lurch, and fled to Messina. In that town he discovered an aged aunt who was sick; the aunt died, and left her money to the Church. Balsamo assumed her family name, added a title of nobility, and was known henceforward as the Count Alessandro Cagliostro.

In Messina he met a mysterious person whom he calls Altotas, and from whom, he says in his Memoir, he learnt much. The following account of the meeting of Balsamo and the stranger is taken from Waite's book: "As he was promenading one day near the jetty at the extremity of the port he encountered an individual singularly habited and possessed of a most remarkable countenance. This person, aged apparently about fifty years, seemed to be an Armenian, though, according to other accounts, he was a Spaniard or Greek. He wore a species of caftan, a silk bonnet, and the extremities of his breeches were concealed in a pair of wide boots. In his left hand he held a parasol, and in his right the end of a cord, to which was attached a graceful Albanian greyhound.... Cagliostro saluted this grotesque being, who bowed slightly, but with satisfied dignity. 'You do not reside in Messina, signor?' he said in Sicilian, but with a marked foreign accent. Cagliostro replied that he was tarrying for a few days, and they began to converse on the beauty of the town and on its advantageous situation, a kind of Oriental imagery individualising the eloquence of the stranger, whose remarks were, moreover, adroitly adorned with a few appropriate compliments."

Although the stranger said he received no one at his house he allowed Cagliostro to visit him. After various mysterious doings the two went off to Egypt, and afterwards to Malta, where they performed many wonderful deeds before the Grand Master, who was much impressed. At Malta Altotas died, or, at anyrate, vanished. Cagliostro then travelled for some time, and was well received by noblemen, ambassadors, and others in high position. At Rome he fell in love with a young and beautiful lady, Lorenza Feliciani, and married her.

Cagliostro used his young wife as a decoy to attract rich and foolish men. He and his wife thrived for a time, and accumulated money and jewels; but a confederate betrayed them, and they fled to Venice, and then wandered for several years in Italy, France, and England. They seem to have made a living by the sale of lotions for the skin, and by practising skilful deceptions.

About the year 1770 Cagliostro began to pose as an alchemist. After another period of wandering he paid a second visit to London and founded a secret society, based on (supposed) Egyptian rites, mingled with those of freemasonry. The suggestion of this society is said to have come from a curious book he picked up on a second-hand stall in London. The society attracted people by the strangeness of its initiatory rites, and the promises of happiness and wellbeing made by its founder to those who joined it. Lodges were established in many countries, many disciples were obtained, great riches were amassed, and Cagliostro flourished exceedingly.

In his Histoire du Merveilleux dans les Temps modernes, Figuier, speaking of Cagliostro about this period of his career, says:

"He proclaimed himself the bearer of the mysteries of Isis and Anubis from the far East.... He obtained numerous and distinguished followers, who on one occasion assembled in great force to hear Joseph Balsamo expound to them the doctrines of Egyptian freemasonry. At this solemn convention he is said to have spoken with overpowering eloquence;... his audience departed in amazement and completely converted to the regenerated and purified masonry. None doubted that he was an initiate of the arcana of nature, as preserved in the temple of Apis at the era when Cambyses belaboured that capricious divinity. From this moment the initiations into the new masonry were numerous, albeit they were limited to the aristocracy of society. There are reasons to believe that the grandees who were deemed worthy of admission paid exceedingly extravagantly for the honour."

Cagliostro posed as a physician, and claimed the power of curing diseases simply by the laying on of hands. He went so far as to assert he had restored to life the dead child of a nobleman in Paris; the discovery that the miracle was effected by substituting a living child for the dead one caused him to flee, laden with spoil, to Warsaw, and then to Strassburg.

Cagliostro entered Strassburg in state, amid an admiring crowd, who regarded him as more than human. Rumour said he had amassed vast riches by the transmutation of base metals into gold. Some people in the crowd said he was the wandering Jew, others that he had been present at the marriage feast of Cana, some asserted he was born before the deluge, and one supposed he might be the devil. The goldsmith whom he had cheated of sixty ounces of gold many years before was in the crowd, and, recognising him, tried to stop the carriage, shouting: "Joseph Balsamo! It is Joseph! Rogue, where are my sixty ounces of gold?" "Cagliostro scarcely deigned to glance at the furious goldsmith; but in the middle of the profound silence which the incident occasioned among the crowd, a voice, apparently in the clouds, uttered with great distinctness the following words: 'Remove this lunatic, who is possessed by infernal spirits.' Some of the spectators fell on their knees, others seized the unfortunate goldsmith, and the brilliant cortege passed on" (Waite).

From Strassburg Cagliostro went to Paris, where he lived in great splendour, curing diseases, making gold and diamonds, mystifying and duping people of all ranks by the splendid ritual and gorgeous feasting of his secret society, and amassing riches. He got entangled in the affair of the Diamond Necklace, and left Paris. Trying to advance his society in Italy he was arrested by the agents of the Inquisition, and imprisoned, then tried, and condemned to death. The sentence was commuted to perpetual imprisonment. After two years in the prison of San Angelo he died at the age of fifty.

CHAPTER IX.

PARACELSUS AND SOME OTHER ALCHEMISTS.

The accounts which have come to us of the men who followed the pursuit of the One Thing are vague, scrappy, and confusing.

Alchemical books abound in quotations from the writings of Geber. Five hundred treatises were attributed to this man during the middle ages, yet we have no certain knowledge of his name, or of the time or place of his birth. Hoefer says he probably lived in the middle of the 8th century, was a native of Mesopotamia, and was named Djabar Al-Konfi. Waite calls him Abou Moussah Djafar al-Sofi. Some of the mediæval adepts spoke of him as the King of India, others called him a Prince of Persia. Most of the Arabian writers on alchemy and medicine, after the 9th century, refer to Geber as their master.

All the MSS. of writings attributed to Geber which have been examined are in Latin, but the library of Leyden is said to possess some works by him written in Arabic. These MSS. contain directions for preparing many metals, salts, acids, oils, etc., and for performing such operations as distillation, cupellation, dissolution, calcination, and the like.

Of the other Arabian alchemists, the most celebrated in the middle ages were Rhasis, Alfarabi, and Avicenna, who are supposed to have lived in the 9th and 10th centuries.

The following story of Alfarabi's powers is taken from Waite's Lives of the Alchemystical Philosophers:—

> "Alfarabi was returning from a pilgrimage to Mecca, when, passing through Syria, he stopped at the Court of the Sultan, and entered his presence, while he was surrounded by numerous sage persons, who were discoursing with the monarch on the sciences. Alfarabi ... presented himself in his travelling attire, and when the Sultan desired he should be seated, with astonishing philosophical freedom he planted himself at the end of the royal sofa. The Prince, aghast at his boldness, called one of his officers, and in a tongue generally unknown commanded him to eject the intruder. The philosopher, however, promptly made answer in the same tongue: 'Oh, Lord, he who acts hastily is liable to hasty repentance.' The Prince was equally

astounded to find himself understood by the stranger as by the manner in which the reply was given. Anxious to know more of his guest he began to question him, and soon discovered that he was acquainted with seventy languages. Problems for discussion were then propounded to the philosophers, who had witnessed the discourteous intrusion with considerable indignation and disgust, but Alfarabi disputed with so much eloquence and vivacity that he reduced all the doctors to silence, and they began writing down his discourse. The Sultan then ordered his musicians to perform for the diversion of the company. When they struck up, the philosopher accompanied them on a lute with such infinite grace and tenderness that he elicited the unmeasured admiration of the whole distinguished assembly. At the request of the Sultan he produced a piece of his own composing, sang it, and accompanied it with great force and spirit to the delight of all his hearers. The air was so sprightly that even the gravest philosopher could not resist dancing, but by another tune he as easily melted them to tears, and then by a soft unobtrusive melody he lulled the whole company to sleep."

The most remarkable of the alchemists was he who is generally known as Paracelsus. He was born about 1493, and died about 1540. It is probable that the place of his birth was Einsiedeln, near Zurich. He claimed relationship with the noble family of Bombast von Hohenheim; but some of his biographers doubt whether he really was connected with that family. His name, or at any rate the name by which he was known, was Aureolus Philippus Theophrastus Bombast von Hohenheim. His father in alchemy, Trimethius, Abbot of Spannheim and then of Wurzburg, who was a theologian, a poet, an astronomer, and a necromancer, named him Paracelsus; this name is taken by some to be a kind of Græco-Latin paraphrase of von Hohenheim (of high lineage), and to mean "belonging to a lofty place"; others say it signifies "greater than Celsus," who was a celebrated Latin writer on medicine of the 1st century. Paracelsus studied at the University of Basle; but, getting into trouble with the authorities, he left the university, and for some years wandered over Europe, supporting himself, according to one account, by "psalm-singing, astrological productions, chiromantic soothsaying, and, it has been said, by necromantic practices." He may have got as far as Constantinople; as a rumour floated about that he received the Stone of Wisdom from an adept in that city. He returned to Basle, and in 1527 delivered lectures with the sanction of the Rector of the university. He made enemies of the physicians by abusing their custom of seeking knowledge only from ancient writers and not from nature; he annoyed the apothecaries by calling their tinctures, decoctions, and extracts, mere soup-messes; and he roused the ire of all learned people by delivering his lectures in German. He was attacked publicly and also anonymously. Of the pamphlets published against him he said, "These vile ribaldries would raise the ire of a turtle-dove." And Paracelsus was no turtle-dove. The following extract from (a translation of) the preface to The Book concerning the Tinctures of the Philosophers written against those Sophists born since the Deluge, shews that his style of writing was abusive, and his opinion of himself, to say the least, not very humble:—

> "From the middle of this age the Monarchy of all the Arts has been at length derived and conferred on me, Theophrastus Paracelsus, Prince of Philosophy and Medicine. For this purpose I have been chosen by God to extinguish and blot out all the phantasies of elaborate and false works, of delusive and presumptuous words, be they the words of Aristotle, Galen, Avicenna, Mesva, or the dogmas of any among their followers. My theory, proceeding as it does from the light of Nature, can never, through its consistency, pass away or be changed; but in the fifty-eighth year after its millennium and a half it will then begin to flourish. The practice at the same time following upon the theory will be proved by wonderful and incredible signs, so as to be open to mechanics and common people, and they will thoroughly understand how firm and immovable is that Paracelsic Art against the triflings of the Sophists; though meanwhile that sophistical science has to have its ineptitude propped up and fortified by papal and imperial privileges.... So then, you wormy and lousy Sophist, since you deem the monarch of Arcana a mere ignorant, fatuous, and prodigal quack, now, in this mid age, I determine in my present treatise to disclose the honourable course of procedure in these matters, the virtues and preparation of the celebrated Tincture of the Philosophers for the use and honour of all who love the truth, and in order that all who despise the true arts may be reduced to poverty."

The turbulent and restless spirit of Paracelsus brought him into open conflict with the authorities of Basle. He fled from that town in 1528, and after many wanderings, he found rest at Salzburg, under the protection of the archbishop. He died at Salzburg in 1541, in his forty-eighth year.

The character and abilities of Paracelsus have been vastly praised by some, and inordinately abused by others. One author says of him: "He lived like a pig, looked like a drover, found his greatest enjoyment in the company of the most dissolute and lowest rabble, and throughout his glorious life he was generally drunk." Another author says: "Probably no physician has grasped his life's task with a purer enthusiasm, or devoted himself more faithfully to it, or more fully maintained the moral worthiness of his calling than did the reformer of Einsiedeln." He certainly seems to have been loved and respected by his pupils and followers, for he is referred to by them as "the noble and beloved monarch," "the German Hemes," and "our dear Preceptor and King of Arts."

There seems no doubt that Paracelsus discovered many facts which became of great importance in chemistry: he prepared the inflammable gas we now call hydrogen, by the reaction between iron filings and oil of vitriol; he distinguished metals from substances which had been classed with metals but lacked the essential metalline character of ductility; he made medicinal preparations of mercury, lead and iron, and introduced many new and powerful drugs, notably laudanum. Paracelsus insisted that medicine is a branch of chemistry, and that the restoration of the body of a patient to a condition of chemical equilibrium is the restoration to health.

Paracelsus trusted in his method; he was endeavouring to substitute direct appeal to nature for appeal to the authority of writers about nature. "After me," he cries, "you Avicenna, Galen, Rhasis, Montagnana and the others. You after me, not I after you. You of Paris, you of Montpellier, you of Swabia, of Meissen and Vienna; you who come from the countries along the Danube and the Rhine; and you, too, from the Islands of the Ocean. Follow me. It is not for me to follow you, for mine is the monarchy." But the work was too arduous, the struggle too unequal. "With few appliances, with no accurate knowledge, with no help from the work of others, without polished and sharpened weapons, and without the skill that comes from long handling of instruments of precision, what could Paracelsus effect in his struggle to wrest her secrets from nature? Of necessity, he grew weary of the task, and tried to construct a universe which should be simpler than that most complex order which refused to yield to his analysis." And so he came back to the universe which man constructs for himself, and exclaimed—

> "Each man has ... all the wisdom and power of the world in himself; he possesses one kind of knowledge as much as another, and he who does not find that which is in him cannot truly say that he does not possess it, but only that he was not capable of successfully seeking for it."

We leave a great genius, with his own words in our ears: "Have no care of my misery, reader; let me bear my burden myself. I have two failings: my poverty and my piety. My poverty was thrown in my face by a Burgomaster who had perhaps only seen doctors attired in silken robes, never basking in tattered rags in the sunshine. So it was decreed I was not a doctor. For my piety I am arraigned by the parsons, for ... I do not at all love those who teach what they do not themselves practise."

CHAPTER X.

SUMMARY OF THE ALCHEMICAL DOCTRINE.—THE REPLACEMENT OF THE THREE PRINCIPLES OF THE ALCHEMISTS BY THE SINGLE PRINCIPLE OF PHLOGISTON.

The Sacred Art, which had its origin and home in Egypt, was very definitely associated with the religious rites, and the theological teaching, recognised by the state. The Egyptian priests were initiated into the mysteries of the divine art: and as the initiated claimed to imitate the work of the deity, the priest was regarded by the ordinary people as something more than a representative, as a mirror, of the divinity. The sacred art of Egypt was transmuted into alchemy by contact with European thought and handicrafts, and the tenets and mysticism of the Catholic Church; and the conception of nature, which was the result of this blending, prevailed from about the 9th until towards the end of the 18th century.

Like its predecessor, alchemy postulated an orderly universe; but alchemy was richer in fantastic details, more picturesquely embroidered, more prodigal of strange fancies, than the sacred art of Egypt.

The alchemist constructed his ordered scheme of nature on the basis of the supposed universality of life. For him, everything lived, and the life of things was threefold. The alchemist thought he recognised the manifestation of life in the form, or body, of a thing, in its soul, and in its spirit. Things might differ much in appearance, in size, taste, smell, and other outward properties, and yet be intimately related, because, according to the alchemist, they were produced from the same principles, they were animated by the same soul. Things might resemble one another closely in their outward properties and yet differ widely in essential features, because, according to the alchemist, they were formed from different elements, in their spiritual properties they were unlike. The alchemists taught that the true transformation, in alchemical language the transmutation, of one thing into another could be effected only by spiritual means acting on the spirit of the thing, because the transmutation consisted essentially in raising the substance to the highest perfection whereof it was capable; the result of this spiritual action might become apparent in the material form of the substance. In attempting to apply such vague conceptions as these, alchemy was obliged to use the language which had been developed for the expression of human emotions and desires, not only for the explanation of the facts it observed, but also for the bare recital of these facts.

The outlook of alchemy on the world outside human beings was essentially anthropomorphic. In the image of man, the alchemist created his universe.

In the times when alchemy was dominant, the divine scheme of creation, and the place given to man in that scheme, were supposed to be thoroughly understood. Everything had its place, designed for it from the beginning, and in that place it remained unless it were forced from it by violent means. A great part of the business of experimental alchemy was to discover the natural position, or condition, of each substance; and the discovery was to be made by interpreting the facts brought to light by observation and experiment by the aid of hypotheses deduced from the general scheme of things which had been formed independently of observation or experiment. Alchemy was a part of magic; for magic interprets and corrects the knowledge gained by the senses by the touchstone of generalisations which have been supplied, partly by the emotions, and partly by extra-human authority, and accepted as necessarily true.

The conception of natural order which regulates the life of the savage is closely related to that which guided the alchemists. The essential features of both are the notion that everything is alive, and the persuasion that things can be radically acted on only by using life as a factor. There is also an intimate connexion between alchemy and witchcraft. Witches were people who were supposed to make an unlawful use of the powers of life; alchemists were often thought to pass beyond what is permitted to the creature, and to encroach on the prerogative of the Creator.

The long duration of alchemy shows that it appealed to some deep-seated want of human beings. Was not that want the necessity for the realisation of order in the universe? Men were unwilling to wait until patient examination of the facts of their own nature, and the facts of nature outside themselves, might lead them to the realisation of the interdependence of all things. They found it easier to evolve a scheme of things from a superficial glance at themselves and their surroundings; naturally they adopted the easier plan. Alchemy was a part of the plan of nature produced by this method. The extraordinary dominancy of such a scheme is testified to by the continued belief in alchemy, although the one experiment, which seems to us to be the crucial experiment of the system, was never accomplished. But it is also to be remembered that the alchemists were acquainted with, and practised, many processes which we should now describe as operations of manufacturing and technical chemistry; and the practical usefulness of these processes bore testimony, of the kind which convinces the plain man, to the justness of their theories.

I have always regarded two facts as most interesting and instructive: that the doctrine of the essential unity of all things, and the simplicity of natural order, was accepted for centuries by many, I think one may say, by most men, as undoubtedly a true presentation of the divine scheme of things; and, secondly, that in more recent times people were quite as certain of the necessary truth of the doctrine, the exact opposite of the alchemical, that the Creator had divided his creation into portions each of which was independent of all the others. Both of these schemes were formed by the same method, by introspection preceding observation; both were overthrown by the same method, by observation and experiment proceeding hand in hand with reasoning. In each case, the humility of science vanquished the conceit of ignorance.

The change from alchemy to chemistry is an admirable example of the change from a theory formed by looking inwards, and then projected on to external facts, to a theory formed by studying facts, and then thinking about them. This change proceeded slowly; it is not possible to name a time when it may be said, here alchemy finishes and chemistry begins. To adapt a saying of one of the alchemists, quoted in a former chapter; alchemy would not easily give up its nature, and fought for its life; but an agent was found strong enough to overcome and kill it, and then that agent also had the power to change the lifeless remains into a new and pure body. The agent was the accurate and imaginative investigation of facts.

The first great step taken in the path which led from alchemy to chemistry was the substitution of one Principle, the Principle of Phlogiston, for the three Principles of salt, sulphur, and mercury. This step was taken by concentrating attention and investigation, by replacing the superficial examination of many diverse phenomena by the more searching study of one class of occurrences. That the field of study should be widened, it was necessary that it should first be narrowed.

Lead, tin, iron, or copper is calcined. The prominent and striking feature of these events is the disappearance of the metal, and the formation of something very unlike it. But the original metal is restored by a second process, which is like the first because it also is a calcination, but seems to differ from the first operation in that the burnt metal is calcined with another substance, with grains of wheat or powdered charcoal. Led thereto by their theory that destruction must precede re-vivification, death must come before resurrection, the alchemists confined their attention to one feature common to all calcinations of metals, and gave a superficial description of these occurrences by classing them together as processes of mortification. Sulphur, wood, wax, oil, and many other things are easily burned: the alchemists said, these things also undergo mortification, they too are killed; but, as "man can restore that which man has destroyed," it must be possible to restore to life the thing which has been mortified. The burnt sulphur, wood, wax, or oil, is not really dead, the alchemists argued; to use the allegory of Paracelsus, they are like young lions which are born dead, and are brought to life by the roaring of their parents: if we make a sufficiently loud noise, if we use the proper means, we shall bring life into what seems to be dead material. As it is the roaring of the parents of the young lions which alone can cause the still-born cubs to live, so it is only by the spiritual agency of life, proceeded the alchemical argument, that life can be brought into the mortified sulphur, wood, wax, and oil.

The alchemical explanation was superficial, theoretical, in the wrong meaning of that word, and unworkable. It was superficial because it overlooked the fact that the primary calcination, the mortification, of the metals, and the other substances, was effected in the air, that is to say, in contact with something different from the thing which was calcined; the explanation was of the kind which people call theoretical, when they wish to condemn an explanation and put it out of court, because it was merely a re-statement of the facts in the language of a theory which had not been deduced from the facts themselves, or from facts like those to be explained, but from what were supposed to be facts without proper investigation, and, if facts, were of a totally different kind from those to which the explanation applied; and lastly, the explanation was unworkable, because it suggested no method whereby its accuracy could be tested, no definite line of investigation which might be pursued.

That great naturalist, the Honourable Robert Boyle (born in 1626, died in 1691), very perseveringly besought those who examined processes of calcination to pay heed to the action of everything which might take part in the processes. He was especially desirous they should consider what part the air might play in calcinations; he spoke of the air as a "menstruum or additament," and said that, in such operations as calcination, "We may well take the freedom to examine ... whether there intervene not a coalition of the parts of the body wrought upon with those of the menstruum, whereby the produced concrete may be judged to result from the union of both."

It was by examining the part played by the air in processes of calcination and burning that men at last became able to give approximately complete descriptions of these processes.

Boyle recognised that the air is not a simple or elementary substance; he spoke of it as "a confused aggregate of effluviums from such differing bodies, that, though they all agree in constituting by their minuteness and various motions one great mass of fluid matter, yet perhaps there is scarce a more heterogeneous body in the world." Clement of Alexandria who lived in the end of the 2nd, and the early part of the 3rd, century A.D., seems to have regarded the air as playing a very important part in combustions; he said—"Airs are divided into two categories; an air for the divine flame, which is the soul; and a material air which is the nourisher of sensible fire, and the basis of combustible matter." Sentences like that I have just quoted are found here and there in the writings of the earlier and later alchemists; now and again we also find statements which may be interpreted, in the light of the fuller knowledge we now have, as indicating at least suspicions that the atmosphere is a mixture of different kinds of air, and that only some of these take part in calcining and burning operations. Those suspicions were confirmed by experiments on the calcination of metals and other substances, conducted in the 17th century by Jean Rey a French physician, and by John Mayow of Oxford. But these observations and the conclusions founded on them, did not bear much fruit until the time of Lavoisier, that is, towards the close of the 18th century. They were overshadowed and put aside by the work of Stahl (1660-1724). Some of the alchemists of the 14th, 15th and 16th centuries taught that combustion and calcination are processes wherein the igneous principle is destroyed, using the word "destroyed" in its alchemical meaning. This description of processes of burning was much more in keeping with the ideas of the time than that given by Boyle, Rey and Mayow. It was adopted by Stahl, and made the basis of a general theory of those changes wherein one substance disappears and another, or others, very unlike it, are produced.

That he might bring into one point of view, and compare the various changes effected by the agency of fire, Stahl invented a new Principle, which he named Phlogiston, and constructed an hypothesis which is generally known as the phlogistic theory. He explained, and applied, this hypothesis in various books, especially in one published at Halle in 1717.

Stahl observed that many substances which differed much from one another in various respects were alike in one respect; they were all combustible. All the combustible substances, he argued, must contain a common principle; he named this supposed principle, phlogiston (from the Greek word phlogistos = burnt, or set on fire). Stahl said that the phlogiston of a combustible thing escapes as the substance burns, and, becoming apparent to the senses, is named fire or flame. The phlogiston in a combustible substance was supposed to be so intimately associated with something else that our senses cannot perceive it; nevertheless, the theory said, it is there; we can see only the escaping phlogiston, we can perceive only the phlogiston which is set free from its combination with other things. The theory thought of phlogiston as imprisoned in the thing which can be burnt, and as itself forming part of the prison; that the prisoner should be set free, the walls of the prison had to be removed; the freeing of the prisoner destroyed the prison. As escaping, or free, phlogiston was called fire, or flame, so the phlogiston in a combustible substance was sometimes called combined fire, or flame in the state of combination. A peculiarity of the strange thing called phlogiston was that it preferred to be concealed in something, hidden, imprisoned, combined; free phlogiston was supposed to be always ready to become combined phlogiston.

The phlogistic theory said that what remains when a substance has been burnt is the original substance deprived of phlogiston; and, therefore, to restore the phlogiston to the product of burning is to re-form the combustible substance. But how is such a restoration of phlogiston to be accomplished? Evidently by heating the burnt thing with something which is very ready to burn. Because, according to the theory, everything which can be burnt contains phlogiston, the more ready a substance is to burn the richer it is in phlogiston; burning is the outrush of phlogiston, phlogiston prefers to be combined with something; therefore, if you mix what remains after burning, with something which is very combustible, and heat the mixture, you are bringing the burnt matter under conditions which are very favourable for the reception of phlogiston by it, for you are bringing it into intimate contact with something from which freedom-hating phlogiston is being forced to escape.

Charcoal, sulphur, phosphorus, oils and fats are easily burnt; these substances were, therefore, chosen for the purpose of changing things which had been burnt into things which could again be burnt; these, and a few other substances like these, were classed together, and called phlogisticating agents.

Very many of the substances which were dealt with by the experimenters of the last quarter of the 17th, and the first half of the 18th, century, were either substances which could be burned, or those which had been produced by burning; hence the phlogistic theory brought into one point of view, compared, and emphasised the similarities between, a great many things which had not been thought of as connected before that theory was promulgated. Moreover, the theory asserted that all combustible, or incinerable, things are composed of phlogiston, and another principle, or, as was often said, another element, which is different in different kinds of combustible substances. The metals, for instance, were said to be composed of phlogiston and an earthy principle or element, which was somewhat different in different metals. The phlogisteans taught that the earthy principle of a metal remains in the form of ash, cinders, or calx, when the metal is calcined, or, as they expressed it, when the metal is deprived of its phlogiston.

The phlogistic theory savoured of alchemy; it postulated an undefined, undefinable, intangible Principle; it said that all combustible substances are formed by the union of this Principle with another, which is sometimes of an earthy character, sometimes of a fatty nature, sometimes highly volatile in habit. Nevertheless, the theory of Stahl was a step away from purely alchemical conceptions towards the accurate description of a very important class of changes. The principle of phlogiston could be recognised by the senses as it was in the act of escaping from a substance; and the other principle of combustible things was scarcely a Principle in the alchemical sense, for, in the case of metals at any rate, it remained when the things which had contained it were burnt, and could be seen, handled, and weighed. To say that metals are composed of phlogiston and an earthy substance, was to express facts in such a language that the expression might be made the basis of experimental inquiry; it was very different from the assertion that metals are produced by the spiritual actions of the three Principles, salt, mercury and sulphur, the first of which is not salt, the second is not mercury, and the third is not sulphur. The followers of Stahl often spoke of metals as composed of phlogiston and an element of an earthy character; this expression also was an advance, from the hazy notion of Element in purely alchemical writings, towards accuracy and fulness of description. An element was now something which could he seen and experimented with; it was no longer a semi-spiritual existence which could not be grasped by the senses.

The phlogistic theory regarded the calcination of a metal as the separation of it into two things, unlike the metal, and unlike each other; one of these things was phlogiston, the other was an earth-like residue. The theory thought of the re-formation of a metal from its calx, that is, the earthy substance which remains after combustion, as the combination of two things to produce one, apparently homogeneous, substance. Metals appeared to the phlogisteans, as they appeared to the alchemists, to be composite substances. Processes of burning were regarded by alchemists and phlogisteans alike, as processes of simplification.

The fact had been noticed and recorded, during the middle ages, that the earth-like matter which remains when a metal is calcined is heavier than the metal itself. From this fact, modern investigators of natural phenomena would draw the conclusion, that calcination of a metal is an addition of something to the metal, not a separation of the metal into different things. It seems impossible to us that a substance should be separated into portions, and one of these parts should weigh as much as, or more than, the whole.

The exact investigation of material changes called chemistry rests on the statement that mass, and mass is practically measured by weight, is the one property of what we call matter, the determination whereof enables us to decide whether a change is a combination, or coalescence, of different things, or a separation of one thing into parts. That any part of a material system can be removed without the weight of the portion which remains being less than the original weight of the whole system, is unthinkable, in the present state of our knowledge of material changes.

But in the 17th century, and throughout most of the 18th, only a few of those who examined changes in the properties of substances paid heed to changes of weight; they had not realised the importance of the property of mass, as measured by weight. The convinced upholder of the phlogistic theory had two answers to the argument, that, because the earth-like product of the calcination of a metal weighs more than the metal itself, therefore the metal cannot have lost something in the process; for, if one portion of what is taken away weighs more than the metal from which it has been separated, it is evident that the weight of the two portions into which the metal is said to have been divided must be considerably greater than the weight of the undivided metal. The upholders of the theory sometimes met the argument by saying, "Of course the calx weighs more than the metal, because phlogiston tends to lighten a body which contains it; and therefore the body weighs more after it has lost phlogiston than it did when the phlogiston formed part of it;" sometimes, and more often, their answer was—"loss or gain of weight is an accident, the essential thing is change of qualities."

If the argument against the separation of a metal into two constituents, by calcination, were answered to-day as it was answered by the upholders of the phlogistic theory, in the middle of the 18th century, the answers would justly be considered inconsequent and ridiculous. But it does not follow that the statements were either far-fetched or absurd at the time they were made. They were expressed in the phraseology of the time; a phraseology, it is true, sadly lacking in consistency, clearness, and appropriateness, but the only language then available for the description of such changes as those which happen when metals are calcined. One might suppose that it must always have sounded ridiculous to say that the weight of a thing can be decreased by adding something to it, that part of a thing weighs more than the whole of it. But the absurdity disappears if it can be admitted that mass, which is measured by weight, may be a property like colour, or taste, or smell; for the colour, taste, or smell of a thing may certainly be made less by adding something else, and the colour, taste, or smell of a thing may also be increased by adding something else. If we did not know that what we call quantity of substance is measured by the property named mass, we might very well accept the proposition that the entrance of phlogiston into a substance decreases the quantity, hence the mass, and, therefore, the weight, of the substance.

Although Stahl and his followers were emerging from the trammels of alchemy, they were still bound by many of the conceptions of that scheme of nature. We have learned, in previous chapters, that the central idea of alchemy was expressed in the saying: "Matter must be deprived of its properties in order to draw out its soul." The properties of substances are everything to the modern chemist—indeed, such words as iron, copper, water, and gold are to him merely convenient expressions for certain definable groups of properties—but the phlogisteans regarded the properties of things, including mass, as of secondary importance; they were still trying to get beneath the properties of a thing, to its hypothetical essence, or substance.

Looking back, we cannot think of phlogiston as a substance, or as a thing, in the modern meanings of these terms as they are used in natural science. Nowadays we think, we are obliged to think, of the sum of the quantities of all the things in the universe as unchanging, and unchangeable by any agency whereof we have definite knowledge. The meaning we give to the word thing rests upon the acceptance of this hypothesis. But the terms substance, thing, properties were used very vaguely a couple of centuries ago; and it would be truly absurd to carry back to that time the meanings which we give to these terms to-day, and then to brand as ridiculous the attempts of the men who studied, then, the same problems which we study now, to express the results of their study in generalisations which employed the terms in question, in what seems to us a loose, vague, and inexact manner.

By asserting, and to some extent experimentally proving, the existence of one principle in many apparently very different substances (or, as would be said to-day, one property common to many substances), the phlogistic theory acted as a very useful means for collecting, and placing in a favourable position for closer inspection, many substances which would probably have remained scattered and detached from one another had this theory not been constructed. A single assumption was made, that all combustible substances are alike in one respect, namely, in containing combined fire, or phlogiston; by the help of this assumption, the theory of phlogiston emphasised the fundamental similarity between all processes of combustion. The theory of phlogiston was extraordinarily simple, compared with the alchemical vagaries which preceded it. Hoefer says, in his Histoire de la Chimie, "If it is true that simplicity is the distinctive character of verity, never was a theory so true as that of Stahl."

The phlogistic theory did more than serve as a means for bringing together many apparently disconnected facts. By concentrating the attention of the students of material changes on one class of events, and giving descriptions of these events without using either of the four alchemical Elements, or the three Principles, Stahl, and those who followed him, did an immense service to the advancement of clear thinking about natural occurrences. The principle of phlogiston was more tangible, and more readily used, than the Salt, Sulphur, and Mercury of the alchemists; and to accustom people to speak of the material substance which remained when a metal, or other combustible substance, was calcined or burnt, as one of the elements of the thing which had been changed, prepared the way for the chemical conception of an element as a definite substance with certain definite properties.

In addition to these advantages, the phlogistic theory was based on experiments, and led to experiments, the results of which proved that the capacity to undergo combustion might be conveyed to an incombustible substance, by causing it to react with some other substance, itself combustible, under definite conditions. The theory thus prepared the way for the representation of a chemical change as an interaction between definite kinds of substances, marked by precise alterations both of properties and composition.The great fault of the theory of phlogiston, considered as a general conception which brings many facts into one point of view, and leads the way to new and exact knowledge, was its looseness, its flexibility. It was very easy to make use of the theory in a broad and general way; by stretching it here, and modifying it there, it seemed to cover all the facts concerning combustion and calcination which were discovered during two generations after the publication of Stahl's books. But many of the subsidiary hypotheses which were required to make the theory cover the new facts were contradictory, or at any rate seemed to be contradictory, of the primary assumptions of the theory. The addition of this ancillary machinery burdened the mechanism of the theory, threw it out of order, and finally made it unworkable. The phlogistic theory was destroyed by its own cumbersomeness.A scientific theory never lasts long if its fundamental assumptions are stated so loosely that they may be easily modified, expanded, contracted, and adjusted to meet the requirements of newly discovered facts. It is true that the theories which have been of the greatest service in science, as summaries of the relations between established facts, and suggestions of lines of investigation, have been stated in terms whose full meaning has gradually unfolded itself. But the foundations of these theories have been at once so rigidly defined and clearly stated as to be incapable of essential modification, and so full of meaning and widely applicable as to cover large classes of facts which were unknown when the theories were constructed. Of the founders of the lasting and expansible theories of natural science, it may be said, that "thoughts beyond their thoughts to those high bards were given."

CHAPTER XI.

THE EXAMINATION OF THE PHENOMENA OF COMBUSTION.

The alchemists thought that the most effectual method of separating a complex substance into more simple substances was to subject it to the action of heat. They were constantly distilling, incinerating, subliming, heating, in order that the spirit, or inner kernel of things, might be obtained. They took for granted that the action of fire was to simplify, and that simplification proceeded whatever might be the nature of the substance which was subjected to this action. Boyle insisted that the effect of heating one substance may be, and often is, essentially different from the effect of heating another substance; and that the behaviour of the same substance when heated, sometimes varies when the conditions are changed. He takes the example of heating sulphur or brimstone: "Exposed to a moderate fire in subliming pots, it rises all into dry, and almost tasteless, flowers; whereas being exposed to a naked fire, it affords store of a saline and fretting liquor." Boyle thought that the action of fire was not necessarily to separate a thing into its principles or elements, but, in most cases, was either to rearrange the parts of the thing, so that new, and it might be, more complex things, were produced, or to form less simple things by the union of the substance with what he called, "the matter of fire." When the product of heating a substance, for example, tin or lead, weighed more than the substance itself, Boyle supposed that the gain in weight was often caused by the "matter of fire" adding itself to the substance which was heated. He commended to the investigation of philosophers this "subtil fluid," which is "able to pierce into the compact and solid bodies of metals, and add something to them that has no despicable weight upon the balance, and is able for a considerable time to continue fixed in the fire." Boyle also drew attention to the possibility of action taking place between a substance which is heated and some other substance, wherewith the original thing may have been mixed. In a word, Boyle showed that the alchemical assumption—fire simplifies—was too simple; and he taught, by precept and example, that the only way of discovering what the action of fire is, on this substance or on that, is to make accurate experiments. "I consider," he says, "that, generally speaking, to render a reason of an effect or phenomenon, is to deduce it from something else in nature more known than itself; and that consequently there may be divers kinds of degrees of explication of the same thing."

Boyle published his experiments and opinions concerning the action of fire on different substances in the seventies of the 17th century; Stahl's books, which laid the foundation of the phlogistic theory, and confirmed the alchemical opinion that the action of fire is essentially a simplifying action, were published about forty years later. But fifty years before Boyle, a French physician, named Jean Rey, had noticed that the calcination of a metal is the production of a more complex, from a less complex substance; and had assigned the increase in weight which accompanies that operation to the attachment of particles of the air to the metal. A few years before the publication of Boyle's work, from which I have quoted, John Mayow, student of Oxford, recounted experiments which led to the conclusion that the air contains two substances, one of which supports combustion and the breathing of animals, while the other extinguishes fire. Mayow called the active component of the atmosphere fiery air; but he was unable to say definitely what becomes of this fiery air when a substance is burnt, although he thought that, in some cases, it probably attaches itself to the burning substances, by which, therefore, it may be said to be fixed. Mayow proved that the air wherein a substance is burnt, or an animal breathes, diminishes in volume during the burning, or the breathing. He tried, without much success, to restore to air that part of it which disappears when combustion, or respiration, proceeds in it.

What happens when a substance is burnt in the air? The alchemists answered this question by asserting that the substance is separated or analysed into things simpler than itself. Boyle said: the process is not necessarily a simplification; it may be, and certainly sometimes is, the formation of something more complicated than the original substance, and when this happens, the process often consists in the fixation of "the matter of fire" by the burning substance. Rey said: calcination, of a metal at anyrate, probably consists in the fixation of particles of air by the substance which is calcined. Mayow answered the question by asserting, on the ground of the results of his experiments, that the substance which is being calcined lays hold of a particular constituent of the air, not the air as a whole.

Now, it is evident that if Mayow's answer was a true description of the process of calcination, or combustion, it should be possible to separate the calcined substance into two different things, one of which would be the thing which was calcined, and the other would be that constituent of the air which had united with the burning, or calcining, substance. It seems clear to us that the one method of proving the accuracy of Mayow's supposition must be, to weigh a definite, combustible, substance—say, a metal; to calcine this in a measured quantity of air; to weigh the product, and to measure the quantity of air which remains; to separate the product of calcination into the original metal, and a kind of air or gas; to prove that the metal thus obtained is the same, and has the same weight, as the metal which was calcined; and to prove that the air or gas obtained from the calcined metal is the same, both in quality and quantity, as the air which disappeared in the process of calcination.

This proof was not forthcoming until about a century after the publication of Mayow's work. The experiments which furnished the proof were rendered possible by a notable discovery made on the 1st of August 1774, by the celebrated Joseph Priestley.

Priestley prepared many "airs" of different kinds: by the actions of acids on metals, by allowing vegetables to decay, by heating beef, mutton, and other animal substances, and by other methods. He says: "Having procured a lens of twelve inches diameter and twenty inches focal distance, I proceeded with great alacrity to examine, by the help of it, what kind of air a great variety of substances, natural and factitious, would yield.... With this apparatus, after a variety of other experiments.... on the 1st of August, 1774, I endeavoured to extract air from mercurius calcinatus per se; and I presently found that, by means of this lens, air was expelled from it very readily. Having got about three or four times as much as the bulk of my materials, I admitted water to it, and found that it was not imbibed by it. But what surprised me more than I can well express was, that a candle burned in this air with a remarkably vigorous flame.... I was utterly at a loss how to account for it."

FIG. XVI.

The apparatus used by Priestley, in his experiments on different kinds of air, is represented in Fig. XVI., which is reduced from an illustration in Priestley's book on Airs.

Priestley had made a discovery which was destined to change Alchemy into Chemistry. But he did not know what his discovery meant. It was reserved for the greatest of all chemists, Antoine Lavoisier, to use the fact stumbled on by Priestley.

After some months Priestley began to think it possible that the new "air" he had obtained from calcined mercury might be fit for respiration. To his surprise he found that a mouse lived in this air much longer than in common air; the new air was evidently better, or purer, than ordinary air. Priestley measured what he called the "goodness" of the new air, by a process of his own devising, and concluded that it was "between four and five times as good as common air."

Priestley was a thorough-going phlogistean. He seems to have been able to describe the results of his experiments only in the language of the phlogistic theory; just as the results of most of the experiments made to-day on the changes of compounds of the element carbon cannot be described by chemists except by making use of the conceptions and the language of the atomic and molecular theory.[6]

The upholder of the phlogistic theory could not think of burning as possible unless there was a suitable receptacle for the phlogiston of the burning substance: when burning occurred in the air, the part played by the air, according to the phlogistic chemist, was to receive the expelled phlogiston; in this sense the air acted as the pabulum, or nourishment, of the burning substance. Inasmuch as substances burned more vigorously and brilliantly in the new air than in common air, Priestley argued that the new air was more ready, more eager, than ordinary air, to receive phlogiston; and, therefore, that the new air contained less phlogiston than ordinary air, or, perhaps, no phlogiston. Arguing thus, Priestley, of course, named the new aeriform substance dephlogisticated air, and thought of it as ordinary air deprived of some, or it might be all, of its phlogiston.

The breathing of animals and the burning of substances were supposed to load the atmosphere with phlogiston. Priestley spoke of the atmosphere as being constantly "vitiated," "rendered noxious," "depraved," or "corrupted" by processes of respiration and combustion; he called those processes whereby the atmosphere is restored to its original condition (or "depurated," as he said), "dephlogisticating processes." As he had obtained his dephlogisticated air by heating the calx of mercury, that is the powder produced by calcining mercury in the air, Priestley was forced to suppose that the calcination of mercury in the air must be a more complex occurrence than merely the expulsion of phlogiston from the mercury: for, if the process consisted only in the expulsion of phlogiston, how could heating what remained produce exceedingly pure ordinary air? It seemed necessary to suppose that not only was phlogiston expelled from mercury during calcination, but that the mercury also imbibed some portion, and that the purest portion, of the surrounding air. Priestley did not, however, go so far as this; he was content to suppose that in some way, which he did not explain, the process of calcination resulted in the loss of phlogiston by the mercury, and the gain, by the dephlogisticated mercury, of the property of yielding exceedingly pure or dephlogisticated air when it was heated very strongly.

Priestley thought of properties in much the same way as the alchemists thought of them, as wrappings, or coverings of an essential something, from which they can be removed and around which they can again be placed. The protean principle of phlogiston was always at hand, and, by skilful management, was ready to adapt itself to any facts. Before the phenomena of combustion could be described accurately, it was necessary to do two things; to ignore the theory of phlogiston, and to weigh and measure all the substances which take part in some selected processes of burning.

Looking back at the attempts made in the past to describe natural events, we are often inclined to exclaim, "Why did investigators bind themselves with the cords of absurd theories; why did they always wear blinkers; why did they look at nature through the distorting mists rising from their own imaginations?" We are too ready to forget the tremendous difficulties which beset the path of him who is seeking accurate knowledge.

"To climb steep hills requires slow pace at first."

Forgetting that the statements wherein the men of science of our own time describe the relations between natural events are, and must be, expressed in terms of some general conception, some theory, of these relations; forgetting that the simplest natural occurrence is so complicated that our powers of description are incapable of expressing it completely and accurately; forgetting the uselessness of disconnected facts; we are inclined to overestimate the importance of our own views of nature's ways, and to underestimate the usefulness of the views of our predecessors. Moreover, as naturalists have not been obliged, in recent times, to make a complete renunciation of any comprehensive theory wherein they had lived and moved for many years, we forget the difficulties of breaking loose from a way of looking at natural events which has become almost as real as the events themselves, of abandoning a language which has expressed the most vividly realised conceptions of generations of investigators, of forming a completely new mental picture of natural occurrences, and developing a completely new language for the expression of those conceptions and these occurrences.

The younger students of natural science of to-day are beginning to forget what their fathers told them of the fierce battle which had to be fought, before the upholders of the Darwinian theory of the origin of species were able to convince those for whom the older view, that species are, and always have been, absolutely distinct, had become a matter of supreme scientific, and even ethical, importance.

A theory which has prevailed for generations in natural science, and has been accepted and used by everyone, can be replaced by a more accurate description of the relations between natural facts, only by the determination, labour, and genius of a man of supreme power. Such a service to science, and humanity, was rendered by Darwin; a like service was done, more than three-quarters of a century before Darwin, by Lavoisier.

Antoine Laurent Lavoisier was born in Paris in 1743. His father, who was a merchant in a good position, gave his son the best education which was then possible, in physical, astronomical, botanical, and chemical science. At the age of twenty-one, Lavoisier gained the prize offered by the Government for devising an effective and economical method of lighting the public streets. From that time until, on the 8th of May 1794, the Government of the Revolution declared, "The Republic has no need of men of science," and the guillotine ended his life, Lavoisier continued his researches in chemistry, geology, physics, and other branches of natural science, and his investigations into the most suitable methods of using the knowledge gained by naturalists for advancing the welfare of the community.

In Chapter VI., I said that when an alchemist boiled water in an open vessel, and obtained a white earthy solid, in place of the water which disappeared, he was producing some sort of experimental proof of the justness of his assertion that water can be changed into earth. Lavoisier began his work on the transformations of matter by demonstrating that this alleged transmutation does not happen; and he did this by weighing the water, the vessel, and the earthy solid.

Lavoisier had constructed for him a pelican of white glass (see Fig. XI., p. 88), with a stopper of glass. He cleaned, dried, and weighed this vessel; then he put into it rain-water which he had distilled eight times; he heated the vessel, removing the stopper from time to time to allow the expanding air to escape, then put in the stopper, allowed the vessel to cool, and weighed very carefully. The difference between the second and the first weighing was the weight of water in the vessel. He then fastened the stopper securely with cement, and kept the apparatus at a temperature about 30° or 40° below that of boiling water, for a hundred and one days. At the end of that time a fine white solid had collected on the bottom of the vessel. Lavoisier removed the cement from the stopper, and weighed the apparatus; the weight was the same as it had been before the heating began. He removed the stopper; air rushed in, with a hissing noise. Lavoisier concluded that air had not penetrated through the apparatus during the process of heating. He then poured out the water, and the solid which had formed in the vessel, set them aside, dried, and weighed the pelican; it had lost 17-4/10 grains. Lavoisier concluded that the solid which had formed in the water was produced by the solvent action of the water on the glass vessel. He argued that if this conclusion was correct, the weight of the solid must be equal to the loss of weight suffered by the vessel; he therefore separated the solid from the water in which it was suspended, dried, and weighed it. The solid weighed 4-9/10 grains. Lavoisier's conclusion seemed to be incorrect; the weight of the solid, which was supposed to be produced by the action of the water on the vessel, was 12 1/2 grains less than the weight of the material removed from the vessel. But some of the material which was removed from the vessel might have remained dissolved in the water: Lavoisier distilled the water, which he had separated from the solid, in a glass vessel, until only a very little remained in the distilling apparatus; he poured this small quantity into a glass basin, and boiled until the whole of the water had disappeared as steam. There remained a white, earthy solid, the weight of which was 15 1/2 grains. Lavoisier had obtained 4 9/10 + 15 1/2 = 20 2/5 grains of solid; the pelican had lost 17 2/5 grains. The difference between these weights, namely, 3 grains, was accounted for by Lavoisier as due to the solvent action of the water on the glass apparatus wherein it had been distilled, and on the glass basin wherein it had been evaporated to dryness.

Lavoisier's experiments proved that when distilled water is heated in a glass vessel, it dissolves some of the material of the vessel, and the white, earthy solid which is obtained by boiling down the water is merely the material which has been removed from the glass vessel. His experiments also proved that the water does not undergo any change during the process; that at the end of the operation it is what it was at the beginning—water, and nothing but water.

By this investigation Lavoisier destroyed part of the experimental basis of alchemy, and established the one and only method by which chemical changes can be investigated; the method wherein constant use is made of the balance.

Lavoisier now turned his attention to the calcination of metals, and particularly the calcination of tin. Boyle supposed that the increase in weight which accompanies the calcination of a metal is due to the fixation of "matter of fire" by the calcining metal; Rey regarded the increase in weight as the result of the combination of the air with the metal; Mayow thought that the atmosphere contains two different kinds of "airs," and one of these unites with the heated metal. Lavoisier proposed to test these suppositions by calcining a weighed quantity of tin in a closed glass vessel, which had been weighed before, and should be weighed after, the calcination. If Boyle's view was correct, the weight of the vessel and the tin would be greater at the end than it was at the beginning of the operation; for "matter of fire" would pass through the vessel and unite with the metal. If there was no change in the total weight of the apparatus and its contents, and if air rushed in when the vessel was opened after the calcination, and the total weight was then greater than at the beginning of the process, it would be necessary to adopt either the supposition of Rey or that of Mayow.

Lavoisier made a series of experiments. The results were these: there was no change in the total weight of the apparatus and its contents; when the vessel was opened after the calcination was finished, air rushed in, and the whole apparatus now weighed more than it did before the vessel was opened; the weight of the air which rushed in was exactly equal to the increase in the weight of the tin produced by the calcination, in other words, the weight of the inrushing air was exactly equal to the difference between the weights of the tin and the calx formed by calcining the tin. Lavoisier concluded that to calcine tin is to cause it to combine with a portion of the air wherein it is calcined. The weighings he made showed that about one-fifth of the whole weight of air in the closed flask wherein he calcined tin had disappeared during the operation.

Other experiments led Lavoisier to suspect that the portion of the air which had united with the tin was different from the portion which had not combined with that metal. He, therefore, set himself to discover whether there are different kinds of "airs" in the atmosphere, and, if there is more than one kind of "air," what is the nature of that "air" which combines with a metal in the process of calcination. He proposed to cause a metallic calx (that is, the substance formed by calcining a metal in the air) to give up the "air" which had been absorbed in its formation, and to compare this "air" with atmospheric air.

About this time Priestley visited Paris, saw Lavoisier, and told him of the new "air" he had obtained by heating calcined mercury. Lavoisier saw the great importance of Priestley's discovery; he repeated Priestley's experiment, and concluded that the air, or gas, which he refers to in his Laboratory Journal as "l'air dephlogistique de M. Priestley" was nothing else than the purest portion of the air we breathe. He prepared this "air" and burned various substances in it. Finding that very many of the products of these combustions had the properties of acids, he gave to the new "air" the name oxygen, which means the acid-producer.

At a later time, Lavoisier devised and conducted an experiment which laid bare the change of composition that happens when mercury is calcined in the air. He calcined a weighed quantity of mercury for many days in a measured volume of air, in an apparatus arranged so that he was able to determine how much of the air disappeared during the process; he collected and weighed the red solid which formed on the surface of the heated mercury; finally he heated this red solid to a high temperature, collected and measured the gas which was given off, and weighed the mercury which was produced. The sum of the weights of the mercury and the gas which were produced by heating the calcined mercury was equal to the weight of the calcined mercury; and the weight of the gas produced by heating the calcined mercury was equal to the weight of the portion of the air which had disappeared during the formation of the calcined mercury. This experiment proved that the calcination of mercury in the air consists in the combination of a constituent of the air with the mercury. Fig. XVII. (reduced from an illustration in Lavoisier's Memoir) represents the apparatus used by Lavoisier. Mayow's supposition was confirmed.

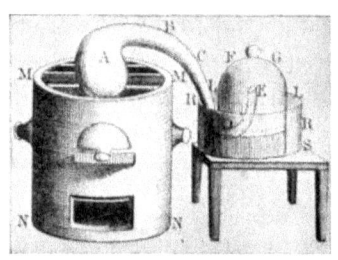

FIG. XVII.

Lavoisier made many more experiments on combustion, and proved that in every case the component of the atmosphere which he had named oxygen combined with the substance, or with some part of the substance, which was burned. By these experiments the theory of Phlogiston was destroyed; and with its destruction, the whole alchemical apparatus of Principles and Elements, Essences and Qualities, Souls and Spirits, disappeared.

CHAPTER XII.

THE RECOGNITION OF CHEMICAL CHANGES AS THE INTERACTIONS OF DEFINITE SUBSTANCES.

The experimental study of combustion made by Lavoisier proved the correctness of that part of Stahl's phlogistic theory which asserted that all processes of combustion are very similar, but also proved that this likeness consists in the combination of a distinct gaseous substance with the material undergoing combustion, and not in the escape therefrom of the Principle of fire, as asserted by the theory of Stahl. After about the year 1790, it was necessary to think of combustions in the air as combinations of a particular gas, or air, with the burning substances, or some portions of them.

This description of processes of burning necessarily led to a comparison of the gaseous constituent of the atmosphere which played so important a part in these processes, with the substances which were burned; it led to the examination of the compositions of many substances, and made it necessary to devise a language whereby these compositions could be stated clearly and consistently.

We have seen, in former chapters, the extreme haziness of the alchemical views of composition, and the connexions between composition and properties. Although Boyle[7] had stated very lucidly what he meant by the composition of a definite substance, about a century before Lavoisier's work on combustion, nevertheless the views of chemists concerning composition remained very vague and incapable of definite expression, until the experimental investigations of Lavoisier enabled him to form a clear mental picture of chemical changes as interactions between definite quantities of distinct substances.

Let us consider some of the work of Lavoisier in this direction. I select his experimental examination of the interactions of metals and acids.

Many experimenters had noticed that gases (or airs, as they were called up till near the end of the 18th century) are generally produced when metals are dissolving in acids. Most of those who noticed this said that the gases came from the dissolving metals; Lavoisier said they were produced by the decomposition of the acids. In order to study the interaction of nitric acid and mercury, Lavoisier caused a weighed quantity of the metal to react with a weighed quantity of the acid, and collected the gas which was produced; when all the metal had dissolved, he evaporated the liquid until a white solid was obtained; he heated this solid until it was changed to the red substance called, at that time, red precipitate, and collected the gas produced. Finally, Lavoisier strongly heated the red precipitate; it changed to a gas, which he collected, and mercury, which he weighed.

The weight of the mercury obtained by Lavoisier at the end of this series of changes was the same, less a few grains, as the weight of the mercury which he had caused to react with the nitric acid. The gas obtained during the solution of the metal in the acid, and during the decomposition of the white solid by heat, was the same as a gas which had been prepared by Priestley and called by him nitrous air; and the gas obtained by heating the red precipitate was found to be oxygen. Lavoisier then mixed measured volumes of oxygen and "nitrous air," standing over water; a red gas was formed, and dissolved in the water, and Lavoisier proved that the water now contained nitric acid.

The conclusions regarding the composition of nitric acid drawn by Lavoisier from these experiments was, that "nitric acid is nothing else than nitrous air, combined with almost its own volume of the purest part of atmospheric air, and a considerable quantity of water."

Lavoisier supposed that the stages in the complete reaction between mercury and nitric acid were these: the withdrawal of oxygen from the acid by the mercury, and the union of the compound of mercury and oxygen thus formed with the constituents of the acid which remained when part of its oxygen was taken away. The removal of oxygen from nitric acid by the mercury produced nitrous air; when the product of the union of the oxide of mercury and the nitric acid deprived of part of its oxygen was heated, more nitrous air was given off, and oxide of mercury remained, and was decomposed, at a higher temperature, into mercury and oxygen.

Lavoisier thought of these reactions as the tearing asunder, by mercury, of nitric acid into definite quantities of its three components, themselves distinct substances, nitrous air, water, and oxygen; and the combination of the mercury with a certain measurable quantity of one of these components, namely, oxygen, followed by the union of this compound of mercury and oxygen with what remained of the components of nitric acid.

Lavoisier had formed a clear, consistent, and suggestive mental picture of chemical changes. He thought of a chemical reaction as always the same under the same conditions, as an action between a fixed and measurable quantity of one substance, having definite and definable properties, with fixed and measurable quantities of other substances, the properties of each of which were definite and definable.

Lavoisier also recognised that certain definite substances could be divided into things simpler than themselves, but that other substances refused to undergo simplification by division into two or more unlike portions. He spoke of the object of chemistry as follows:—[8] "In submitting to experiments the different substances found in nature, chemistry seeks to decompose these substances, and to get them into such conditions that their various components may be examined separately. Chemistry advances to its end by dividing, sub-dividing, and again sub-dividing, and we do not know what will be the limits of such operations. We cannot be certain that what we regard as simple to-day is indeed simple; all we can say is, that such a substance is the actual term whereat chemical analysis has arrived, and that with our present knowledge we cannot sub-divide it."

In these words Lavoisier defines the chemical conception of elements; since his time an element is "the actual term whereat chemical analysis has arrived," it is that which "with our present knowledge we cannot sub-divide"; and, as a working hypothesis, the notion of element has no wider meaning than this. I have already quoted Boyle's statement that by elements he meant "certain primitive and simple bodies ... not made of any other bodies, or of one another." Boyle was still slightly restrained by the alchemical atmosphere around him; he was still inclined to say, "this must be the way nature works, she must begin with certain substances which are absolutely simple." Lavoisier had thrown off all the trammels which hindered the alchemists from making rigorous experimental investigations. If one may judge from his writings, he had not struggled to free himself from these trammels, he had not slowly emerged from the quagmires of alchemy, and painfully gained firmer ground; the extraordinary clearness and directness of his mental vision had led him straight to the very heart of the problems of chemistry, and enabled him not only calmly to ignore all the machinery of Elements, Principles, Essences, and the like, which the alchemists had constructed so laboriously, but also to construct, in place of that mechanism which hindered inquiry, genuine scientific hypotheses which directed inquiry, and were themselves altered by the results of the experiments they had suggested. Lavoisier made these great advances by applying himself to the minute and exhaustive examination of a few cases of chemical change, and endeavouring to account for everything which took part in the processes he studied, by weighing or measuring each distinct substance which was present when the change began, and each which was present when the change was finished. He did not make haphazard experiments; he had a method, a system; he used hypotheses, and he used them rightly. "Systems in physics," Lavoisier writes, "are but the proper instruments for helping the feebleness of our senses. Properly speaking, they are methods of approximation which put us on the track of solving problems; they are the hypotheses which, successively modified, corrected, and changed, by experience, ought to conduct us, some day, by the method of exclusions and eliminations, to the knowledge of the true laws of nature."

In a memoir wherein he is considering the production of carbonic acid and alcohol by the fermentation of fruit-juice, Lavoisier says, "It is evident that we must know the nature and composition of the substances which can be fermented and the products of fermentation; for nothing is created, either in the operations of art or in those of nature; and it may be laid down that the quantity of material present at the beginning of every operation is the same as the quantity present at the end, that the quality and quantity of the principles[9] are the same, and that nothing happens save certain changes, certain modifications. On this principle is based the whole art of experimenting in chemistry; in all chemical experiments we must suppose that there is a true equality between the principles[10] of the substances which are examined and those which are obtained from them by analysis."

If Lavoisier's memoirs are examined closely, it is seen that at the very beginning of his chemical inquiries he assumed the accuracy, and the universal application, of the generalisation "nothing is created, either in the operations of art or in those of nature." Naturalists had been feeling their way for centuries towards such a generalisation as this; it had been in the air for many generations; sometimes it was almost realised by this or that investigator, then it escaped for long periods. Lavoisier seems to have realised, by what we call intuition, that however great and astonishing may be the changes in the properties of the substances which mutually react, there is no change in the total quantity of material. Not only did Lavoisier realise and act on this principle, he also measured quantities of substances by the one practical method, namely, by weighing; and by doing this he showed chemists the only road along which they could advance towards a genuine knowledge of material changes. The generalisation expressed by Lavoisier in the words I have quoted is now known as the law of the conservation of mass; it is generally stated in some such form as this:—the sum of the masses of all the homogeneous substances which take part in a chemical (or physical) change does not itself change. The science of chemistry rests on this law; every quantitative analysis assumes the accuracy, and is a proof of the validity, of it.[11] By accepting the accuracy of this generalisation, and using it in every experiment, Lavoisier was able to form a clear mental picture of a chemical change as the separation and combination of homogeneous substances; for, by using the balance, he was able to follow each substance through the maze of changes, to determine when it united with other substances, and when it separated into substances simpler than itself.

CHAPTER XIII.

THE CHEMICAL ELEMENTS CONTRASTED WITH THE ALCHEMICAL PRINCIPLES.

It was known to many observers in the later years of the 17th century that the product of the calcination of a metal weighs more than the metal; but it was still possible, at that time, to assert that this fact is of no importance to one who is seeking to give an accurate description of the process of calcination. Weight, which measures mass or quantity of substance, was thought of, in these days, as a property like colour, taste, or smell, a property which was sometimes decreased, and sometimes increased, by adding one substance to another. Students of natural occurrences were, however, feeling their way towards the recognition of some property of substances which did not change in the haphazard way wherein most properties seemed to alter. Lavoisier reached this property at one bound. By his experimental investigations, he taught that, however greatly the properties of one substance may be masked, or altered, by adding another substance to it, yet the property we call mass, and measure by weight, is not affected by these changes; for Lavoisier showed, that the mass of the product of the union of two substances is always exactly the sum of the masses of these two substances, and the sum of the masses of the substances whereinto one substance is divided is always exactly equal to that mass of the substance which is divided.

For the undefined, ever-changing, protean essence, or soul, of a thing which the alchemists thought of as hidden by wrappings of properties, the exact investigations of Lavoisier, and those of others who worked on the same lines as he, substituted this definite, fixed, unmodifiable property of mass. Lavoisier, and those who followed in his footsteps, also did away with the alchemical notion of the existence of an essential substratum, independent of changes in those properties of a substance which can be observed by the senses. For the experimental researches of these men obliged naturalists to recognise, that a change in the properties of a definite, homogeneous substance, such as pure water, pure chalk, or pure sulphur, is accompanied (or, as we generally say, is caused) by the formation of a new substance or substances; and this formation, this apparent creation, of new material, is effected, either by the addition of something to the original substance, or by the separation of it into portions which are unlike it, and unlike one another. If the change is a combination, or coalescence, of two things into one, then the mass, and hence the weight, of the product is equal to the sum of those masses, and hence those weights, of the things which have united to form it; if the change is a separation of one distinct substance into several substances, then the sum of the masses, and hence the weights, of the products is equal to that mass, and hence that weight, of the substance which has been separated.

Consider the word water, and the substance represented by this word. In Chapter IV., I gave illustrations of the different meanings which have been given to this word; it is sometimes used to represent a material substance, sometimes a quality more or less characteristic of that substance, and sometimes a process to which that substance, and many others like it, may be subjected. But when the word water is used with a definite and exact meaning, it is a succinct expression for a certain group, or collocation, of measurable properties which are always found together, and is, therefore, thought of as a distinct substance. This substance can be separated into two other substances very unlike it, and can be formed by causing these to unite. One hundred parts, by weight, of pure water are always formed by the union of 11.11 parts of hydrogen, and 88.89 parts of oxygen, and can be separated into these quantities of those substances. When water is formed by the union of hydrogen and oxygen, in the ratio of 11.11 parts by weight of the former to 88.89 of the latter, the properties of the two substances which coalesce to form it disappear, except their masses. It is customary to say that water contains hydrogen and oxygen; but this expression is scarcely an accurate description of the facts. What we call substances are known to us only by their properties, that is, the ways wherein they act on our senses. Hydrogen has certain definite properties, oxygen has other definite properties, and the properties of water are perfectly distinct from those of either of the substances which it is said to contain. It is, therefore, somewhat misleading to say that water contains substances the properties whereof, except their masses, disappeared at the moment when they united and water was produced. Nevertheless we are forced to think of water as, in a sense, containing hydrogen and oxygen. For, one of the properties of hydrogen is its power to coalesce, or combine, with oxygen to form water, and one of the properties of oxygen is its ability to unite with hydrogen to form water; and these properties of those substances cannot be recognised, or even suspected, unless certain definite quantities of the two substances are brought together under certain definite conditions. The properties which characterise hydrogen, and those which characterise oxygen, when these things are separated from all other substances, can be determined and measured in terms of the similar properties of some other substance taken as a standard. These two distinct substances disappear when they are brought into contact, under the proper conditions, and something (water) is obtained whose properties are very unlike those of hydrogen or oxygen; this new thing can be caused to disappear, and hydrogen and oxygen are again produced. This cycle of changes can be repeated as often as we please; the quantities of hydrogen and oxygen which are obtained when we choose to stop the process are exactly the same as the quantities of those substances which disappeared in the first operation whereby water was produced. Hence, water is an intimate union of hydrogen and oxygen; and, in this sense, water may be said to contain hydrogen and oxygen.

The alchemist would have said, the properties of hydrogen and oxygen are destroyed when these things unite to form water, but the essence, or substratum, of each remains. The chemist says, you cannot discover all the properties of hydrogen and oxygen by examining these substances apart from one another, for one of the most important properties of either is manifested only when the two mutually react: the formation of water is not the destruction of the properties of hydrogen and oxygen and the revelation of their essential substrata, it is rather the manifestation of a property of each which cannot be discovered except by causing the union of both.

There was, then, a certain degree of accuracy in the alchemical description of the processes we now call chemical changes, as being the removal of the outer properties of the things which react, and the manifestation of their essential substance. But there is a vast difference between this description and the chemical presentment of these processes as reactions between definite and measurable quantities of elements, or compounds, or both, resulting in the re-distribution, of the elements, or the separation of the compounds into their elements, and the formation of new compounds by the re-combination of these elements.

Let us contrast the two descriptions somewhat more fully.

The alchemist wished to effect the transmutation of one substance into another; he despaired of the possibility of separating the Elements whereof the substance might be formed, but he thought he could manipulate what he called the virtues of the Elements by a judicious use of some or all of the three Principles, which he named Sulphur, Salt, and Mercury. He could not state in definite language what he meant by these Principles; they were states, conditions, or qualities, of classes of substances, which could not be defined. The directions the alchemist was able to give to those who sought to effect the change of one thing into another were these. Firstly, to remove those properties which characterised the thing to be changed, and leave only the properties which it shared with other things like it; secondly, to destroy the properties which the thing to be changed possessed in common with certain other things; thirdly, to commingle the Essence of the thing with the Essence of something else, in due proportion and under proper conditions; and, finally, to hope for the best, keep a clear head, and maintain a sense of virtue.

If he who was about to attempt the transmutation inquired how he was to destroy the specific properties, and the class properties, of the thing he proposed to change, and by what methods he was to obtain its Essence, and cause that Essence to produce the new thing, he would be told to travel along "the road which was followed by the Great Architect of the Universe in the creation of the world." And if he demanded more detailed directions, he would be informed that the substance wherewith his experiments began must first be mortified, then dissolved, then conjoined, then putrefied, then congealed, then cibated, then sublimed, and fermented, and, finally, exalted. He would, moreover, be warned that in all these operations he must use, not things which he could touch, handle, and weigh, but the virtues, the lives, the souls, of such things.

When the student of chemistry desires to effect the transformation of one definite substance into another, he is told to determine, by quantitative experiments, what are the elements, and what the quantities of these elements, which compose the compound which he proposes to change, and the compound into which he proposes to change it; and he is given working definitions of the words element and compound. If the compound he desires to produce is found to be composed of elements different from those which form the compound wherewith his operations begin, he is directed to bring about a reaction, or a series of reactions, between the compound which is to be changed, and some other collocation of elements the composition of which is known to be such that it can supply the new elements which are needed for the production of the new compound.

Since Lavoisier realised, for himself, and those who were to come after him, the meaning of the terms element and compound, we may say that chemists have been able to form a mental picture of the change from one definite substance to another, which is clear, suggestive, and consistent, because it is an approximately accurate description of the facts discovered by careful and penetrative investigations. This presentment of the change has been substituted for the alchemical conception, which was an attempt to express what introspection and reasoning on the results of superficial investigations, guided by specious analogies, suggested ought to be the facts.

Lavoisier was the man who made possible the more accurate, and more far-reaching, description of the changes which result in the production of substances very unlike those which are changed; and he did this by experimentally analysing the conceptions of the element and the compound, giving definite and workable meanings to these conceptions, and establishing, on an experimental foundation, the generalisation that the sum of the quantities of the substances which take part in any change is itself unchanged.

A chemical element was thought of by Lavoisier as "the actual term whereat analysis has arrived," a definite substance "which we cannot subdivide with our present knowledge," but not necessarily a substance which will never be divided. A compound was thought of by him as a definite substance which is always produced by the union of the same quantities of the same elements, and can be separated into the same quantities of the same elements.

These conceptions were amplified and made more full of meaning by the work of many who came after Lavoisier, notably by John Dalton, who was born in 1766 and died in 1844.

In Chapter I., I gave a sketch of the atomic theory of the Greek thinkers. The founder of that theory, who flourished about 500 B.C., said that every substance is a collocation of a vast number of minute particles, which are unchangeable, indestructible, and impenetrable, and are therefore properly called atoms; that the differences which are observed between the qualities of things are due to differences in the numbers, sizes, shapes, positions, and movements of atoms, and that the process which occurs when one substance is apparently destroyed and another is produced in its place, is nothing more than a rearrangement of atoms.

The supposition that changes in the properties of substances are connected with changes in the numbers, movements, and arrangements of different kinds of minute particles, was used in a general way by many naturalists of the 17th and 18th centuries; but Dalton was the first to show that the data obtained by the analyses of compounds make it possible to determine the relative weights of the atoms of the elements.

Dalton used the word atom to denote the smallest particle of an element, or a compound, which exhibits the properties characteristic of that element or compound. He supposed that the atoms of an element are never divided in any of the reactions of that element, but the atoms of a compound are often separated into the atoms of the elements whereof the compound is composed. Apparently without knowing that the supposition had been made more than two thousand years before his time, Dalton was led by his study of the composition and properties of the atmosphere to assume that the atoms of different substances, whether elements or compounds, are of different sizes and have different weights. He assumed that when two elements unite to form only one compound, the atom of that compound has the simplest possible composition, is formed by the union of a single atom of each element. Dalton knew only one compound of hydrogen and nitrogen, namely, ammonia. Analyses of this compound show that it is composed of one part by weight of hydrogen and 4.66 parts by weight of nitrogen. Dalton said one atom of hydrogen combines with one atom of nitrogen to form an atom of ammonia; hence an atom of nitrogen is 4.66 times heavier than an atom of hydrogen; in other words, if the atomic weight of hydrogen is taken as unity, the atomic weight of nitrogen is expressed by the number 4.66. Dalton referred the atomic weights of the elements to the atomic weight of hydrogen as unity, because hydrogen is lighter than any other substance; hence the numbers which tell how much heavier the atoms of the elements are than an atom of hydrogen are always greater than one, are always positive numbers.

When two elements unite in different proportions, by weight, to form more than one compound, Dalton supposed that (in most cases at any rate) one of the compounds is formed by the union of a single atom of each element; the next compound is formed by the union of one atom of the element which is present in smaller quantity with two, three, or more, atoms of the other element, and the next compound is formed by the union of one atom of the first element with a larger number (always, necessarily, a whole number) of atoms of the other element than is contained in the second compound; and so on. From this assumption, and the Daltonian conception of the atom, it follows that the quantities by weight of one element which are found to unite with one and the same weight of another element must always be expressible as whole multiples of one number. For if two elements, A and B, form a compound, that compound is formed, by supposition, of one atom of A and one atom of B; if more of B is added, at least one atom of B must be added; however much of B is added the quantity must be a whole number of atoms; and as every atom of B is the same in all respects as every other atom of B, the weights of B added to a constant weight of A must be whole multiples of the atomic weight of B.

The facts which were available in Dalton's time confirmed this deduction from the atomic theory within the limits of experimental errors; and the facts which have been established since Dalton's time are completely in keeping with the deduction. Take, for instance, three compounds of the elements nitrogen and oxygen. That one of the three which contains least oxygen is composed of 63.64 per cent. of nitrogen, and 36.36 per cent. of oxygen; if the atomic weight of nitrogen is taken to be 4.66, which is the weight of nitrogen that combines with one part by weight of hydrogen, then the weight of oxygen combined with 4.66 of nitrogen is 2.66 (63.64:36.36 = 4.66:2.66). The weights of oxygen which combine with 4.66 parts by weight of nitrogen to form the second and third compounds, respectively, must be whole multiples of 2.66; these weights are 5.32 and 10.64. Now 5.32 = 2.66 x 2, and 10.64 = 2.66 x 4. Hence, the quantities by weight of oxygen which combine with one and the same weight of nitrogen are such that two of these quantities are whole multiples of the third quantity.

Dalton's application of the Greek atomic theory to the facts established by the analyses of compounds enabled him to attach to each element a number which he called the atomic weight of the element, and to summarise all the facts concerning the compositions of compounds in the statement, that the elements combine in the ratios of their atomic weights, or in the ratios of whole multiples of their atomic weights. All the investigations which have been made into the compositions of compounds, since Dalton's time, have confirmed the generalisation which followed from Dalton's application of the atomic theory.

Even if the theory of atoms were abandoned, the generalisation would remain, as an accurate and exact statement of facts which hold good in every chemical change, that a number can be attached to each element, and the weights of the elements which combine are in the ratios of these numbers, or whole multiples of these numbers.

Since chemists realised the meaning of Dalton's book, published in 1808, and entitled, A New System of Chemical Philosophy, elements have been regarded as distinct and definite substances, which have not been divided into parts different from themselves, and unite with each other in definite quantities by weight which can be accurately expressed as whole multiples of certain fixed quantities; and compounds have been regarded as distinct and definite substances which are formed by the union of, and can be separated into, quantities of various elements which are expressible by certain fixed numbers or whole multiples thereof. These descriptions of elements and compounds are expressions of actual facts. They enable chemists to state the compositions of all the compounds which are, or can be, formed by the union of any elements. For example, let A, B, C, and D represent four elements, and also certain definite weights of these elements, then the compositions of all the compounds which can be formed by the union of these elements are expressed by the scheme $A_n B_m C_p D_q$, where m n p and q are whole numbers.

These descriptions of elements and compounds also enable chemists to form a clear picture to themselves of any chemical change. They think of a chemical change as being; (1) a union of those weights of two, or more, elements which are expressed by the numbers attached to these elements, or by whole multiples of these numbers; or (2) a union of such weights of two, or more, compounds as can be expressed by certain numbers or by whole multiples of these numbers; or (3) a reaction between elements and compounds, or between compounds and compounds, resulting in the redistribution of the elements concerned, in such a way that the complete change of composition can be expressed by using the numbers, or whole multiples of the numbers, attached to the elements.

How different is this conception of a change wherein substances are formed, entirely unlike those things which react to form them, from the alchemical presentment of such a process! The alchemist spoke of stripping off the outer properties of the thing to be changed, and, by operating spiritually on the soul which was thus laid bare, inducing the essential virtue of the substance to exhibit its powers of transmutation. But he was unable to give definite meanings to the expressions which he used, he was unable to think clearly about the transformations which he tried to accomplish. The chemist discards the machinery of virtues, souls, and powers. It is true that he substitutes a machinery of minute particles; but this machinery is merely a means of thinking clearly and consistently about the changes which he studies. The alchemist thought, vaguely, of substance as something underlying, and independent of, properties; the chemist uses the expression, this or that substance, as a convenient way of presenting and reasoning about certain groups of properties. It seems to me that if we think of matter as something more than properties recognised by the senses, we are going back on the road which leads to the confusion of the alchemical times.

The alchemists expressed their conceptions in what seems to us a crude, inconsistent, and very undescriptive language. Chemists use a language which is certainly symbolical, but also intelligible, and on the whole fairly descriptive of the facts.

A name is given to each elementary substance, that is, each substance which has not been decomposed; the name generally expresses some characteristic property of the substance, or tells something about its origin or the place of its discovery. The names of compounds are formed by putting together the names of the elements which combine to produce them; and the relative quantities of these elements are indicated either by the use of Latin or Greek prefixes, or by variations in the terminal syllables of the names of the elements.

CHAPTER XIV.

THE MODERN FORM OF THE ALCHEMICAL QUEST OF THE ONE THING.

The study of the properties of the elements shows that these substances fall into groups, the members of each of which are like one another, and form compounds which are similar. The examination of the properties and compositions of compounds has shown that similarity of properties is always accompanied by similarity of composition. Hence, the fact that certain elements are very closely allied in their properties suggests that these elements may also be allied in their composition. Now, to speak of the composition of an element is to think of the element as formed by the union of at least two different substances; it implies the supposition that some elements at any rate are really compounds.

The fact that there is a very definite connexion between the values of the atomic weights, and the properties, of the elements, lends some support to the hypothesis that the substances we call, and are obliged at present to call, elements, may have been formed from one, or a few, distinct substances, by some process of progressive change. If the elements are considered in the order of increasing atomic weights, from hydrogen, whose atomic weight is taken as unity because it is the lightest substance known, to uranium, an atom of which is 240 times heavier than an atom of hydrogen, it is found that the elements fall into periods, and the properties of those in one period vary from element to element, in a way which is, broadly and on the whole, like the variation of the properties of those in other periods. This fact suggests the supposition—it might be more accurate to say the speculation—that the elements mark the stable points in a process of change, which has not proceeded continuously from a very simple substance to a very complex one, but has repeated itself, with certain variations, again and again. If such a process has occurred, we might reasonably expect to find substances exhibiting only minute differences in their properties, differences so slight as to make it impossible to assign the substances, definitely and certainly, either to the class of elements or to that of compounds. We find exactly such substances among what are called the rare earths. There are earth-like substances which exhibit no differences of chemical properties, and yet show minute differences in the characters of the light which they emit when they are raised to a very high temperature.

The results of analysis by the spectroscope of the light emitted by certain elements at different temperatures may be reasonably interpreted by supposing that these elements are separated into simpler substances by the action on them of very large quantities of thermal energy. The spectrum of the light emitted by glowing iron heated by a Bunsen flame (say, at 1200° C. = about 2200° F.) shows a few lines and flutings; when iron is heated in an electric arc (say, to 3500° C. = about 6300° F.) the spectrum shows some two thousand lines; at the higher temperature produced by the electric spark-discharge, the spectrum shows only a few lines. As a guide to further investigation, we may provisionally infer from these facts that iron is changed at very high temperatures into substances simpler than itself.

Sir Norman Lockyer's study of the spectra of the light from stars has shown that the light from those stars which are presumably the hottest, judging by the general character of their spectra, reveals the presence of a very small number of chemical elements; and that the number of spectral lines, and, therefore, the number of elements, increases as we pass from the hottest to cooler stars. At each stage of the change from the hottest to cooler stars certain substances disappear and certain other substances take their places. It may be supposed, as a suggestive hypothesis, that the lowering of stellar temperature is accompanied by the formation, from simpler forms of matter, of such elements as iron, calcium, manganese, and other metals.

In the year 1896, the French chemist Becquerel discovered the fact that salts of the metal uranium, the atomic weight of which is 240, and is greater than that of any other element, emit rays which cause electrified bodies to lose their electric charges, and act on photographic plates that are wrapped in sheets of black paper, or in thin sheets of other substances which stop rays of light. The radio-activity of salts of uranium was proved not to be increased or diminished when these salts had been shielded for five years from the action of light by keeping them in leaden boxes. Shortly after Becquerel's discovery, experiments proved that salts of the rare metal thorium are radio-active. This discovery was followed by Madame Curie's demonstration of the fact that certain specimens of pitchblende, a mineral which contains compounds of uranium and of many other metals, are extremely radio-active, and by the separation from pitchblende, by Monsieur and Madame Curie, of new substances much more radio-active than compounds of uranium or of thorium. The new substances were proved to be compounds chemically very similar to salts of barium. Their compositions were determined on the supposition that they were salts of an unknown metal closely allied to barium. Because of the great radio-activity of the compounds, the hypothetical metal of them was named Radium. At a later time, radium was isolated by Madame Curie. It is described by her as a white, hard, metal-like solid, which reacts with water at the ordinary temperature, as barium does.

Since the discovery of radium compounds, many radio-active substances have been isolated. Only exceedingly minute quantities of any of them have been obtained. The quantities of substances used in experiments on radio-activity are so small that they escape the ordinary methods of measurement, and are scarcely amenable to the ordinary processes of the chemical laboratory. Fortunately, radio-activity can be detected and measured by electrical methods of extraordinary fineness, methods the delicacy of which very much more exceeds that of spectroscopic methods than the sensitiveness of these surpasses that of ordinary chemical analysis.

At the time of the discovery of radio-activity, about seventy-five substances were called elements; in other words, about seventy-five different substances were known to chemists, none of which had been separated into unlike parts, none of which had been made by the coalescence of unlike substances. Compounds of only two of these substances, uranium and thorium, are radio-active. Radio-activity is a very remarkable phenomenon. So far as we know at present, radio-activity is not a property of the substances which form almost the whole of the rocks, the waters, and the atmosphere of the earth; it is not a property of the materials which constitute living organisms. It is a property of some thirty substances—of course, the number may be increased—a few of which are found widely distributed in rocks and waters, but none of which is found anywhere except in extraordinarily minute quantity. Radium is the most abundant of these substances; but only a very few grains of radium chloride can be obtained from a couple of tons of pitchblende.

In Chapter X. of The Story of the Chemical Elements I have given a short account of the outstanding phenomena of radio-activity; for the present purpose it will suffice to state a few facts of fundamental importance.

Radio-active substances are stores of energy, some of which is constantly escaping from them; they are constantly changing without external compulsion, and are constantly radiating energy: all explosives are storehouses of energy which, or part of which, can be obtained from them; but the liberation of their energy must be started by some kind of external shock. When an explosive substance has exploded, its existence as an explosive is finished; the products of the explosion are substances from which energy cannot be obtained: when a radio-active substance has exploded, it explodes again, and again, and again; a time comes, sooner or later, when it has changed into substances that are useless as sources of energy. The disintegration of an explosive, started by an external force, is generally completed in a fraction of a second; change of condition changes the rate of explosion: the "half-life period" of each radio-active substance is a constant characteristic of it; if a gram of radium were kept for about 1800 years, half of it would have changed into radio-inactive substances.

Conditions may be arranged so that an explosive remains unchanged—wet gun-cotton is not exploded by a shock which would start the explosion of dry gun-cotton—in other words, the explosion of an explosive can be regulated: the explosive changes of a radio-active substance, which are accompanied by the radiation of energy, cannot be regulated; they proceed spontaneously in a regular and definable manner which is not influenced by any external conditions—such as great change of temperature, presence or absence of other substances—so far as these conditions have been made the subject of experiment: the amount of activity of a radio-active substance has not been increased or diminished by any process to which the substance has been subjected. Explosives are manufactured articles; explosiveness is a property of certain arrangements of certain quantities of certain elements: so far as experiments have gone, it has not been found possible to add the property of radio-activity to an inactive substance, or to remove the property of radio-activity from an active substance; the cessation of the radio-activity of an active substance is accompanied by the disappearance of the substance, and the production of inactive bodies altogether unlike the original active body.

Radio-active substances are constantly giving off energy in the form of heat, sending forth rays which have definite and remarkable properties, and producing gaseous emanations which are very unstable, and change, some very rapidly, some less rapidly, into other substances, and emit rays which are generally the same as the rays emitted by the parent substance. In briefly considering these three phenomena, I shall choose radium compounds as representative of the class of radio-active substances.

Radium compounds spontaneously give off energy in the form of heat. A quantity of radium chloride which contains 1 gram of radium continuously gives out, per hour, a quantity of heat sufficient to raise the temperature of 1 gram of water through 100° C., or 100 grams of water through 1° C. The heat given out by 1 gram of radium during twenty-four hours would raise the temperature of 2400 grams of water through 1° C.; in one year the temperature of 876,000 grams of water would be raised through 1° C.; and in 1800 years, which is approximately the half-life period of radium, the temperature of 1,576,800 kilograms of water would be raised through 1° C. These results may be expressed by saying that if 1 gram (about 15 grains) of radium were kept until half of it had changed into inactive substances, and if the heat spontaneously produced during the changes which occurred were caused to act on water, that quantity of heat would raise the temperature of about 15½ tons of water from its freezing- to its boiling-point.

Radium compounds send forth three kinds of rays, distinguished as alpha, beta, and gamma rays. Experiments have made it extremely probable that the α-rays are streams of very minute particles, somewhat heavier than atoms of hydrogen, moving at the rate of about 18,000 miles per second; and that the β-rays are streams of much more minute particles, the mass of each of which is about one one-thousandth of the mass of an atom of hydrogen, moving about ten times more rapidly than the α-particles, that is, moving at the rate of about 180,000 miles per second. The γ-rays are probably pulsations of the ether, the medium supposed to fill space. The emission of α-rays by radium is accompanied by the production of the inert elementary gas, helium; therefore, the α-rays are, or quickly change into, rapidly moving particles of helium. The particles which constitute the β-rays carry electric charges; these electrified particles, each approximately a thousand times lighter than an atom of hydrogen, moving nearly as rapidly as the pulsations of the ether which we call light, are named electrons. The rays from radium compounds discharge electrified bodies, ionise gases, that is, cause them to conduct electricity, act on photographic plates, and produce profound changes in living organisms.

The radium emanation is a gas about 111 times heavier than hydrogen; to this gas Sir William Ramsay has given the name niton. The gas has been condensed to a colourless liquid, and frozen to an opaque solid which glows like a minute arc-light. Radium emanation gives off α-particles, that is, very rapidly moving atoms of helium, and deposits exceedingly minute quantities of a solid, radio-active substance known as radium A. The change of the emanation into helium and radium A proceeds fairly rapidly: the half-life period of the emanation is a little less than four days. This change is attended by the liberation of much energy.

The only satisfactory mental picture which the facts allow us to form, at present, of the emission of β-rays from radium compounds is that which represents these rays as streams of electrons, that is, particles, each about a thousand times lighter than an atom of hydrogen, each carrying an electric charge, and moving at the rate of about 180,000 miles per second, that is, nearly as rapidly as light. When an electric discharge is passed from a plate of metal, arranged as the kathode, to a metallic wire arranged as the anode, both sealed through the walls of a glass tube or bulb from which almost the whole of the air has been extracted, rays proceed from the kathode, in a direction at right angles thereto, and, striking the glass in the neighbourhood of the anode, produce a green phosphorescence. Facts have been gradually accumulated which force us to think of these kathode rays as streams of very rapidly moving electrons, that is, as streams of extraordinarily minute electrically charged particles identical with the particles which form the β-rays emitted by compounds of radium.

The phenomena of radio-activity, and also the phenomena of the kathode rays, have obliged us to refine our machinery of minute particles by including therein particles at least a thousand times lighter than atoms of hydrogen. The term electron was suggested, a good many years ago, by Dr Johnstone Stoney, for the unit charge of electricity which is carried by an atom of hydrogen when hydrogen atoms move in a liquid or gas under the directing influence of the electric current. Some chemists speak of the electrons, which are the β-rays from radium, and the kathode rays produced in almost vacuous tubes, as non-material particles of electricity. Non-material means devoid of mass. The method by which approximate determinations have been made of the charges on electrons consists in measuring the ratio between the charges and the masses of these particles. If the results of the determinations are accepted, electrons are not devoid of mass. Electrons must be thought of as material particles differing from other minute material particles in the extraordinary smallness of their masses, in the identity of their properties, including their mass, in their always carrying electric charges, and in the vast velocity of their motion. We must think of an electron either as a unit charge of electricity one property of which is its minute mass, or as a material particle having an extremely small mass and carrying a unit charge of electricity: the two mental pictures are almost, if not quite, identical.

Electrons are produced by sending an electric discharge through a glass bulb containing a minute quantity of air or other gas, using metallic plates or wires as kathode and anode. Experiments have shown that the electrons are identical in all their properties, whatever metal is used to form the kathode and anode, and of whatever gas there is a minute quantity in the bulb. The conclusion must be drawn that identical electrons are constituents of, or are produced from, very different kinds of chemical elements. As the facts about kathode rays, and the facts of radio-activity are (at present) inexplicable except on the supposition that these phenomena are exhibited by particles of extraordinary minuteness, and as the smallest particles with which chemists are concerned in their everyday work are the atoms of the elements, we seem obliged to think of many kinds of atoms as structures, not as homogeneous bodies. We seem obliged to think of atoms as very minute material particles, which either normally are, or under definite conditions may be, associated with electrically charged particles very much lighter than themselves, all of which are identical, whatever be the atoms with which they are associated or from which they are produced.

In their study of different kinds of matter, chemists have found it very helpful to place in one class those substances which they have not been able to separate into unlike parts. They have distinguished this class of substances from other substances, and have named them elements. The expression chemical elements is merely a summary of certain observed facts. For many centuries chemists have worked with a conceptual machinery based on the notion that matter has a grained structure. For more than a hundred years they have been accustomed to think of atoms as the ultimate particles with which they have had to deal. Working with this order-producing instrument, they have regarded the properties of elements as properties of the atoms, or of groups of a few of the atoms, of these substances. That they might think clearly and suggestively about the properties of elements, and connect these with other chemical facts, they have translated the language of sense-perceptions into the language of thought, and, for properties of those substances which have not been decomposed, have used the more fertile expression atomic properties. When a chemist thinks of an atom, he thinks of the minutest particle of one of the substances which have the class-mark have-not-been-decomposed, and the class-name element. The chemist does not call these substances elements because he has been forced to regard the minute particles of them as undivided, much less because he thinks of these particles as indivisible; his mental picture of their structure as an atomic structure formed itself from the fact that they had not been decomposed. The formation of the class element followed necessarily from observed facts, and has been justified by the usefulness of it as an instrument for forwarding accurate knowledge. The conception of the elementary atom as a particle which had not been decomposed followed from many observed facts besides those concerning elements, and has been justified by the usefulness of it as an instrument for forwarding accurate knowledge. Investigations proved radio-activity to be a property of the very minute particles of certain substances, and each radio-active substance to have characteristic properties, among which were certain of those that belong to elements, and to some extent are characteristic of elements. Evidently, the simplest way for a chemist to think about radio-activity was to think of it as an atomic property; hence, as atomic properties had always been regarded, in the last analysis, as properties of elements, it was natural to place the radio-active substances in the class elements, provided that one forgot for the time that these substances have not the class-mark have-not-been-decomposed.

As the facts of radio-activity led to the conclusion that some of the minute particles of radio-active substances are constantly disintegrating, and as these substances had been labelled elements, it seemed probable, or at least possible, that the other bodies which chemists have long called elements are not true elements, but are merely more stable collocations of particles than the substances which are classed as compounds. As compounds can be changed into certain other compounds, although not into any other compounds, a way seemed to be opening which might lead to the transformation of some elements into some other elements.

The probability that one element might be changed into another was increased by the demonstration of the connexions between uranium and radium. The metal uranium has been classed with the elements since it was isolated in 1840. In 1896, Becquerel found that compounds of uranium, and also the metal itself, are radio-active. In the light of what is now known about radio-activity, it is necessary to suppose that some of the minute particles of uranium emit particles lighter than themselves, and change into some substance, or substances, different from uranium; in other words, it is necessary to suppose that some particles of uranium are spontaneously disintegrating. This supposition is confirmed by the fact, experimentally proved, that uranium emits α-rays, that is, atoms of helium, and produces a substance known as uranium X. Uranium X is itself radio-active; it emits β-rays, that is, it gives off electrons. Inasmuch as all minerals which contain compounds of uranium contain compounds of radium also, it is probable that radium is one of the disintegration-products of uranium. The rate of decay of radium may be roughly expressed by saying that, if a quantity of radium were kept for ten thousand years, only about one per cent. of the original quantity would then remain unchanged. Even if it were assumed that at a remote time the earth's crust contained considerable quantities of radium compounds, it is certain that they would have completely disappeared long ago, had not compounds of radium been reproduced from other materials. Again, the most likely hypothesis is that compounds of radium are being produced from compounds of uranium.

Uranium is a substance which, after being rightly classed with the elements for more than half a century, because it had not been separated into unlike parts, must now be classed with the radium-like substances which disintegrate spontaneously, although it differs from other radio-active substances in that its rate of change is almost infinitively slower than that of any of them, except thorium.[12] Thorium, a very rare metal, is the second of the seventy-five or eighty elements known when radio-activity was discovered, which has been found to undergo spontaneous disintegration with the emission of rays. The rate of change of thorium is considerably slower than that of uranium.[13] None of the other substances placed in the class of elements is radio-active.

On p. 192 I said, that when the radio-active substances had been labelled elements, the facts of radio-activity led some chemists to the conclusion that the other bodies which had for long been called by this class-name, or at any rate some of these bodies, are perhaps not true elements, but are merely more stable collocations of particles than the substances called compounds. It seems to me that this reasoning rests on an unscientific use of the term element; it rests on giving to that class-name the meaning, substances asserted to be undecomposable. A line of demarcation is drawn between elements, meaning thereby forms of matter said to be undecomposable but probably capable of separation into unlike parts, and true elements, meaning thereby groups of identical undecomposable particles. If one names the radio-active substances elements, one is placing in this class substances which are specially characterised by a property the direct opposite of that the possession of which by other substances was the reason for the formation of the class. To do this may be ingenious; it is certainly not scientific.

Since the time of Lavoisier, since the last decade of the eighteenth century, careful chemists have meant by an element a substance which has not been separated into unlike parts, and they have not meant more than that. The term element has been used by accurate thinkers as a useful class-mark which connotes a property—the property of not having been decomposed—common to all substances placed in the class, and differentiating them from all other substances. Whenever chemists have thought of elements as the ultimate kinds of matter with which the physical world is constructed—and they have occasionally so thought and written—they have fallen into quagmires of confusion.

Of course, the elements may, some day, be separated into unlike parts. The facts of radio-activity certainly suggest some kind of inorganic evolution. Whether the elements are decomposed is to be determined by experimental inquiry, remembering always that no number of failures to simplify them will justify the assertion that they cannot be simplified. Chemistry neither asserts or denies the decomposability of the elements. At present, we have to recognise the existence of extremely small quantities, widely distributed in rocks and waters, of some thirty substances, the minute particles of which are constantly emitting streams of more minute, identical particles that carry with them very large quantities of energy, all of which thirty substances are characterised, and are differentiated from all other classes of substances wherewith chemistry is concerned, by their spontaneous mutability, and each is characterised by its special rate of change and by the nature of the products of its mutations. We have now to think of the minute particles of two of the seventy-five or eighty substances which until the other day had not been decomposed, and were therefore justly called elements, as very slowly emitting streams of minuter particles and producing characteristic products of their disintegration. And we have to think of some eighty substances as particular kinds of matter, at present properly called elements, because they are characterised, and differentiated from all other substances, by the fact that none of them has been separated into unlike parts.

The study of radio-activity has introduced into chemistry and physics a new order of minute particles. Dalton made the atom a beacon-light which revealed to chemists paths that led them to wider and more accurate knowledge. Avogadro illuminated chemical, and also physical, ways by his conception of the molecule as a stable, although separable, group of atoms with particular properties different from those of the atoms which constituted it. The work of many investigators has made the old paths clearer, and has shown to chemists and physicists ways they had not seen before, by forcing them to think of, and to make use of, a third kind of material particles that are endowed with the extraordinary property of radio-activity. Dalton often said: "Thou knowest thou canst not cut an atom"; but the fact that he applied the term atom to the small particles of compounds proves that he had escaped the danger of logically defining the atom, the danger of thinking of it as a particle which never can be cut. The molecule of Avogadro has always been a decomposable particle. The peculiarity of the new kind of particles, the particles of radio-active bodies, is, not that they can be separated into unlike parts by the action of external forces, but that they are constantly separating of their own accord into unlike parts, and that their spontaneous disintegration is accompanied by the production of energy, the quantity of which is enormous in comparison with the minuteness of the material specks which are the carriers of it.

The continued study of the properties of the minute particles of radio-active substances—a new name is needed for those most mutable of material grains—must lead to discoveries of great moment for chemistry and physics. That study has already thrown much light on the phenomena of electric conductivity; it has given us the electron, a particle at least a thousand times lighter than an atom of hydrogen; it has shown us that identical electrons are given off by, or are separated from, different kinds of elementary atoms, under definable conditions; it has revealed unlooked-for sources of energy; it has opened, and begun the elucidation of, a new department of physical science; it has suggested a new way of attacking the old problem of the alchemists, the problem of the transmutation of the elements.

The minute particles of two of the substances for many years classed as elements give off electrons; uranium and thorium are radio-active. Electrons are produced by sending an electric discharge through very small traces of different gases, using electrodes of different metals. Electrons are also produced by exposing various metals to the action of ultra-violet light, and by raising the temperature of various metals to incandescence. Electrons are always identical, whatever be their source. Three questions suggest themselves. Can the atoms of all the elements be caused to give off electrons? Are electrons normal constituents of all elementary atoms? Are elementary atoms collocations of electrons? These questions are included in the demand—Is it possible "to imagine a model which has in it the potentiality of explaining" radio-activity and other allied phenomena, as well as all other chemical and physical properties of elements and compounds? These questions are answerable by experimental investigation, and only by experimental investigation. If experimental inquiry leads to affirmative answers to the questions, we shall have to think of atoms as structures of particles much lighter than themselves; we shall have to think of the atoms of all kinds of substances, however much the substances differ chemically and physically, as collocations of identical particles; we shall have to think of the properties of atoms as conditioned, in our final analysis, by the number and the arrangement of their constitutive electrons. Now, if a large probability were established in favour of the view that different atoms are collocations of different numbers of identical particles, or of equal numbers of differently arranged identical particles, we should have a guide which might lead to methods whereby one collocation of particles could be formed from another collocation of the same particles, a guide which might lead to methods whereby one element could be transformed into another element.

To attempt "to imagine a model which has in it the potentiality of explaining" radio-activity, the production of kathode rays, and the other chemical and physical properties of elements and compounds, might indeed seem to be a hopeless undertaking. A beginning has been made in the mental construction of such a model by Professor Sir J.J. Thomson. To attempt a description of his reasoning and his results is beyond the scope of this book.[14]

The facts that the emanation from radium compounds spontaneously gives off very large quantities of energy, and that the emanation can easily be brought into contact with substances on which it is desired to do work, suggested to Sir William Ramsay that the transformation of compounds of one element into compounds of another element might possibly be effected by enclosing a solution of a compound along with radium emanation in a sealed tube, and leaving the arrangement to itself. Under these conditions, the molecules of the compound would be constantly bombarded by a vast number of electrons shot forth at enormous velocities from the emanation. The notion was that the molecules of the compound would break down under the bombardment, and that the atoms so produced might be knocked into simpler groups of particles—in other words, changed into other atoms—by the terrific, silent shocks of the electrons fired at them incessantly by the disintegrating emanation. Sir William Ramsay regards his experimental results as establishing a large probability in favour of the assertion that compounds of copper were transformed into compounds of lithium and sodium, and compounds of thorium, of cerium, and of certain other rare metals, into compounds of carbon. The experimental evidence in favour of this statement has not been accepted by chemists as conclusive. A way has, however, been opened which may lead to discoveries of great moment.

Let us suppose that the transformation of one element into another element or into other elements has been accomplished. Let us suppose that the conception of elementary atoms as very stable arrangements of many identical particles, from about a thousand to about a quarter of a million times lighter than the atoms, has been justified by crucial experiments. Let us suppose that the conception of the minute grains of radio-active substances as particular but constantly changing arrangements of the same identical particles, stable groups of which are the atoms of the elements, has been firmly established. One result of the establishment of the electronic conception of atomic structure would be an increase of our wonder at the complexity of nature's ways, and an increase of our wonder that it should be possible to substitute a simple, almost rigid, mechanical machinery for the ever-changing flow of experience, and, by the use of that mental mechanism, not only to explain very many phenomena of vast complexity, but also to predict occurrences of similar entanglement and to verify these predictions.

The results which have been obtained in the examination of radio-activity, of kathode rays, of spectra at different temperatures, and of phenomena allied to these, bring again into prominence the ancient problem of the structure of what we call matter. Is matter fundamentally homogeneous or heterogeneous? Chemistry studies the relations between the changes of composition and the changes of properties which happen simultaneously in material systems. The burning fire of wood, coal, or gas; the preparation of food to excite and to satisfy the appetite; the change of minerals into the iron, steel, copper, brass, lead, tin, lighting burning and lubricating oils, dye-stuffs and drugs of commerce; the change of the skins, wool, and hair of animals, and of the seeds and fibres of plants, into clothing for human beings; the manufacture from rags, grass, or wood of a material fitted to receive and to preserve the symbols of human hopes, fears, aspirations, love and hate, pity and aversion; the strange and most delicate processes which, happening without cessation, in plants and animals and men, maintain that balanced equilibrium which we call life; and, when the silver cord is being loosed and the bowl broken at the cistern, the awful changes which herald the approach of death; not only the growing grass in midsummer meadows, not only the coming of autumn "in dyed garments, travelling in the glory of his apparel," but also the opening buds, the pleasant scents, the tender colours which stir our hearts in "the spring time, the only pretty ring time, when birds do sing, ding-a*—dong-ding": these, and a thousand other changes have all their aspects which it is the business of the chemist to investigate. Confronted with so vast a multitude of never-ceasing changes, and bidden to find order there, if he can—bidden, rather compelled by that imperious command which forces the human mind to seek unity in variety, and, if need be, to create a cosmos from a chaos; no wonder that the early chemists jumped at the notion that there must be, that there is, some One Thing, some Universal Essence, which binds into an orderly whole the perplexing phenomena of nature, some Water of Paradise which is for the healing of all disorder, some "Well at the World's End," a draught whereof shall bring peace and calm security.

The alchemists set forth on the quest. Their quest was barren. They made the great mistake of fashioning The One Thing, The Essence, The Water of Paradise, from their own imaginings of what nature ought to be. In their own likeness they created their goal, and the road to it. If we are to understand nature, they cried, her ways must be simple; therefore, her ways are simple. Chemists are people of a humbler heart. Their reward has been greater than the alchemists dreamed. By selecting a few instances of material changes, and studying these with painful care, they have gradually elaborated a general conception of all those transformations wherein substances are produced unlike those by the interaction of which they are formed. That general conception is now both widening and becoming more definite. To-day, chemists see a way opening before them which they reasonably hope will lead them to a finer, a more far-reaching, a more suggestive, at once a more complex and a simpler conception of material changes than any of those which have guided them in the past.

FOOTNOTES

1 Most of the quotations from alchemical writings, in this book, are taken from a series of translations, published in 1893-94, under the supervision of Mr A.E. Waite.

2 The quotations from Lucretius are taken from Munro's translation (4th Edition, 1886).

3 See the chapter Molecular Architecture in the Story of the Chemical Elements.

4 The author I am quoting had said—"Nature is divided into four 'places' in which she brings forth all things that appear and that are in the shade; and according to the good or bad quality of the 'place,' she brings forth good or bad things.... It is most important for us to know her 'places' ... in order that we may join things together according to Nature."

5 The account of the life of Cagliostro is much condensed from Mr A.E. Waite's Lives of the Alchemystical Philosophers.

6 I have given numerous illustrations of the truth of this statement in the book, in this series, entitled The Story of the Wanderings of Atoms.

7 Boyle said, in 1689, "I mean by elements ... certain primitive and simple, or perfectly unmixed bodies; which not being made of any other bodies, or of one another, are the ingredients of which all those called perfectly mixt bodies are immediately compounded, and into which they are ultimately resolved."

8 I have given a free rendering of Lavoisier's words.

9, 10 Lavoisier uses the word principle, here and elsewhere, to mean a definite homogeneous substance; he uses it as synonymous with the more modern terms element and compound.

11 I have considered the law of the conservation of mass in some detail in Chapter IV. of The Story of the Chemical Elements.

12 The life-period of uranium is probably about eight thousand million years.

13 The life-period of thorium is possibly about forty thousand million years.

14 The subject is discussed in Sir J.J. Thomson's Electricity and Matter.